Table of Contents

Sections: **Day:**

(Answer Key in Back)

ISBN: 978-1-63578-324-7 UK Edition

Current contact information can be found at
www.HumbleMath.com
www.LibroStudioLLC.com

Cover image credit: Helen Stebakov/Shutterstock.com

Day 1

2-Step Problems

Name: ____________________

Score:

① $13 + 7 \times 3$

13 + 21

34

② $9 \times 2 - 13$

③ $32 \div 8 + 4$

④ $29 - 24 \div 2$

⑤ $20 \div 4 \times 9$

⑥ $15 - 7 \times 2$

⑦ $14 + 8 \div 2$

⑧ $45 \div 9 \times 2$

⑨ $30 - 20 \div 5$

⑩ $24 + 6 \times 6$

Day 2

2-Step Problems

Name: ____________________

Score:

① 45 - 40 ÷ 8

② 28 + 8 ÷ 4

③ 3 + 2 x 9

④ 6 x 4 - 7

⑤ 10 ÷ 2 x 5

⑥ 10 - 2 x 4

⑦ 16 + 4 ÷ 2

⑧ 24 ÷ 3 x 2

⑨ 12 - 4 ÷ 4

⑩ 2 + 2 x 3

Day 3

2-Step Problems

Name: ____________________

① 5 + 5 x 6

② 3 x 5 - 4

③ 49 ÷ 7 + 10

④ 19 - 21 ÷ 3

⑤ 15 ÷ 15 x 5

⑥ 17 - 18 ÷ 3

⑦ 33 + 1 x 1

⑧ 9 ÷ 3 x 6

⑨ 50 - 35 x 1

⑩ 13 + 12 ÷ 4

Day 4

2-Step Problems

Name: ____________________

Score:

① 28 - 16 ÷ 8

② 16 x 4 + 20

③ 4 + 1 - 2

④ 18 ÷ 9 x 2

⑤ 28 - 3 x 9

⑥ 2 + 7 x 18

⑦ 9 x 3 + 5

⑧ 16 ÷ 8 - 2

⑨ 20 - 15 ÷ 3

⑩ 40 + 2 - 28

Day 5

2-Step Problems

Name: ____________________

① 30 x 4 + 5

② 5 - 3 ÷ 3

③ 8 + 1 x 10

④ 16 ÷ 8 + 27

⑤ 50 x 3 ÷ 15

⑥ 44 - 40 ÷ 4

⑦ 17 + 20 - 7

⑧ 16 x 8 - 50

⑨ 5 - 2 x 1

⑩ 6 + 20 - 20

Day 6

3-Step Problems

Name: ______________

Score:

① 6 + 10 - 18 ÷ 3

② 13 +7 x 15 ÷ 5

③ 28 - 12 x 2 + 6

④ 32 ÷ 8 x 5 + 4

⑤ 26 + 8 x 24 ÷ 6

⑥ 20 ÷ 4 x 11 + 3

⑦ 4 + 8 ÷ 2 x 3

⑧ 14 + 14 ÷ 7 - 2

⑨ 3 x 12 + 4 x 10

⑩ 10 x 2 - 16 ÷ 4

Day 7

3-Step Problems

Name: ____________________

Score:

① $3 \times 5 - 15 \div 5$

② $48 - 40 + 2 \times 4$

③ $17 - 6 \div 3 \times 1$

④ $5 + 37 \times 24 - 23$

⑤ $22 - 3 \times 5 + 20$

⑥ $3 + 9 \div 3 \times 3$

⑦ $9 \times 6 - 38 + 5$

⑧ $6 - 3 + 8 \times 2$

⑨ $7 \times 5 - 3 \div 3$

⑩ $1 + 49 \times 4 \div 4$

Day 8

3-Step Problems

Name: ____________________

Score:

① 28 + 6 x 3 - 1

② 9 x 6 ÷ 2 + 4

③ 36 ÷ 6 + 15 x 2

④ 11 - 5 x 2 + 3

⑤ 22 + 18 - 40 ÷ 8

⑥ 35 - 5 ÷ 1 x 7

⑦ 50 ÷ 50 x 20 + 20

⑧ 7 x 16 + 21 - 11

⑨ 9 + 2 - 1 x 8

⑩ 13 - 5 x 0 + 34

Day 9

3-Step Problems

Name: ____________________

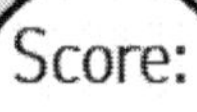

Score:

① $7 \times 13 - 14 \div 7$

② $27 + 0 \times 6 - 6$

③ $41 - 20 \div 10 \times 9$

④ $8 \div 8 + 3 \times 3$

⑤ $2 \times 22 + 17 - 16$

⑥ $5 + 39 - 36 \div 9$

⑦ $44 - 2 \div 2 + 4$

⑧ $12 \div 6 \times 37 - 33$

⑨ $5 + 3 - 15 \div 5$

⑩ $6 \times 10 + 23 - 5$

Day 10

3-Step Problems

Name: ____________________

Score:

① 36 ÷ 9 + 11 x 3

② 18 - 2 x 6 + 48

③ 12 x 1 - 6 + 2

④ 4 + 14 ÷ 1 x 8

⑤ 9 - 3 x 3 + 6

⑥ 36 ÷ 12 + 24 - 11

⑦ 17 x 3 - 19 + 25

⑧ 16 + 16 ÷ 16 - 8

⑨ 41 - 12 + 1 x 0

⑩ 2 x 10 - 18 ÷ 2

Day 11

3-Step Problems

Name: ____________________

Score:

① $1 + 17 \times 12 - 8$

② $35 \div 5 - 2 \times 3$

③ $6 \times 1 + 6 \div 2$

④ $26 - 12 \div 6 + 4$

⑤ $43 + 8 - 16 \times 3$

⑥ $9 \div 9 \times 2 - 2$

⑦ $1 \times 23 + 6 - 1$

⑧ $14 - 3 \times 2 + 17$

⑨ $50 \div 25 - 1 \times 1$

⑩ $13 - 15 \div 3 + 5$

Day 12

3-Step Problems

Name: ____________________

Score:

① $7 \times 7 - 34 \div 2$

② $24 - 4 \div 2 \times 1$

③ $3 + 9 \times 9 - 3$

④ $45 \div 5 + 0 - 0$

⑤ $20 - 4 \div 2 \times 8$

⑥ $11 \times 6 - 26 + 6$

⑦ $8 + 19 \times 21 \div 7$

⑧ $14 \div 14 + 3 - 2$

⑨ $8 - 4 \div 1 + 1$

⑩ $26 \times 2 - 37 + 8$

Day 13

3-Step Problems

Name: ____________________

① 50 - 26 + 8 ÷ 4

② 28 + 18 ÷ 9 - 28

③ 36 ÷ 6 - 2 + 1

④ 1 x 9 + 13 - 15

⑤ 13 - 8 x 1 + 2

⑥ 41 + 6 - 8 x 3

⑦ 1 x 25 ÷ 5 + 1

⑧ 33 ÷ 3 + 27 x 3

⑨ 41 + 16 - 16 ÷ 4

⑩ 19 - 7 x 2 + 7

Day 14

3-Step Problems

Name: ____________________

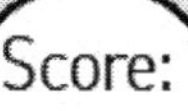

① 6 x 9 - 50 + 2

② 19 - 18 x 3 ÷ 3

③ 42 + 49 ÷ 7 - 40

④ 38 ÷ 2 - 1 x 5

⑤ 8 x 7 + 16 ÷ 8

⑥ 15 - 14 x 1 + 4

⑦ 10 - 4 ÷ 1 x 1

⑧ 17 + 6 - 30 ÷ 5

⑨ 11 x 2 + 25 - 22

⑩ 44 ÷ 4 x 13 + 7

Day 15

3-Step Problems

Name: ____________________

Score:

① 35 - 14 ÷ 2 x 5

② 12 + 37 x 24 ÷ 8

③ 4 x 26 - 50 + 40

④ 9 ÷ 9 + 5 - 2

⑤ 0 + 1 x 1 ÷ 1

⑥ 6 x 9 ÷ 9 + 25

⑦ 6 x 6 - 13 + 8

⑧ 27 ÷ 3 + 24 x 4

⑨ 48 - 28 + 5 x 43

⑩ 15 + 5 x 7 - 7

Day 16

Exponents

Name: ____________________

Score:

① $6 + 7^2 - 18 \div 3$

② $24 + 3^2 \times 4 \div 2$

③ $21 - 6 \times 3 + 8^2$

④ $6^2 \div 3 \times 2 + 2^2$

⑤ $31 + 4^2 \div 2 \times 6$

⑥ $20 \div 4 \times 3^2 - 5^2$

⑦ $230 - 8 \div 2^2 \times 7$

⑧ $9^2 + 24 \div 8 - 2$

⑨ $3 \times 5 + 4^2 \times 10$

⑩ $5^2 \times 2 - 40 \div 5$

Day 17

Exponents

Name: ____________________

Score:

① $40 \div 10 + 1 \times 4^2$

② $5^2 \times 4 \div 20 + 50$

③ $31 + 6 - 9^2 \div 3$

④ $49 - 2^2 \times 8 + 2$

⑤ $1 \times 2^2 + 3^2 - 2$

⑥ $5^2 + 16 \div 16 - 10$

⑦ $18 \div 9 - 1^2 + 1$

⑧ $2^2 \times 2^2 - 6 + 5$

⑨ $7 + 6 \times 6 \div 3^2$

⑩ $32 - 12 + 6^2 \times 1$

Day 18

Exponents

Name: ________________

Score:

① $8^2 - 9 \times 7 + 8^2$

② $2^2 + 3^2 - 18 \div 6$

③ $2 \times 5^2 + 41 - 34$

④ $16 \div 4^2 + 1 \times 7^2$

⑤ $14 + 39 - 6^2 \div 3$

⑥ $48 - 18 \div 3^2 + 1$

⑦ $2 \times 2 + 20 - 4^2$

⑧ $27 \div 3^2 - 1 \times 2$

⑨ $25 - 10 \div 2 + 1^2$

⑩ $1^2 + 5 \times 2^2 - 1$

Day 19

Exponents

Name: ____________________

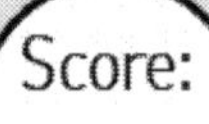

① $26 + 7^2 - 37 \times 1^2$

② $85 - 15 \times 2^2 + 2$

③ $49 \div 7^2 + 6^2 - 22$

④ $5^2 \times 3 \div 15 - 5$

⑤ $20 - 4 \times 5 + 4^2$

⑥ $11 + 2^2 - 35 \div 7$

⑦ $6^2 \times 1 \div 2 - 4^2$

⑧ $54 - 27 + 3^2 \times 3$

⑨ $14 + 1^2 - 18 \div 3^2$

⑩ $36 \div 12 \times 5^2 - 50$

Day 20

Exponents

Name: ____________________

Score:

① $50 - 7^2 \times 1 + 4$

② $2^2 \times 7 + 5^2 - 20$

③ $27 + 2 - 7^2 \div 7$

④ $10^2 \div 10 \times 1 + 5^2$

⑤ $6 \times 3^2 + 13 - 12$

⑥ $32 - 8 \div 1^2 \times 4$

⑦ $2^2 + 38 - 2 \times 4^2$

⑧ $7 \times 8 + 6^2 \div 6^2$

⑨ $19 - 16 \div 1^2 + 2$

⑩ $18 \div 3^2 \times 25 - 24$

Day 21

Parentheses

Name: ___________________

Score:

① $(6 + 10) \div 8 - 1$

② $4 + 4 \times (15 - 5)$

③ $25 - (6 + 2^2 \times 3)$

④ $(4^2 - 9) \times (8 + 2)$

⑤ $(22 + 8)2 \div 6$

⑥ $20 \div (2 \times 2) + 3$

⑦ $(4 + 20) \div (3^2 - 3)$

⑧ $14 \div (9 + 7 - 2)$

⑨ $(3 + 5 \times 4)10$

⑩ $32 \div 8(6 + 4)$

Day 22

Parentheses

Name: ____________________

Score:

① $6 + 6^2 \div (9 - 3)$

② $2^2 + 4(30 - 5^2)$

③ $100 - (3 \times 5 + 7^2)$

④ $(4^2 - 14)^2 (10 - 3^2)$

⑤ $(2 + 8)3^2 \div 3$

⑥ $5^2 \div (7 - 2) + 14$

⑦ $150 \div (8 - 3)^2 \times 2$

⑧ $18 \div (6^2 - 5 \times 6)$

⑨ $(8^2 + 3 \times 4)10$

⑩ $(8 + 4^2) \div (5 + 3)$

Day 23

Parentheses

Name: ________________

Score:

① $(2 \times 4^2)\ (27 - 26)$

② $9\ (3 + 7^2) - 5$

③ $2^2 + 6\ (25 \div 5^2)$

④ $(3^2 + 7 \times 3^2)\ 2$

⑤ $(47 - 46) + 6^2 \times 6$

⑥ $(10^2 + 10) \div (10 + 1^2)$

⑦ $49 + (45 \div 3^2 - 2^2)$

⑧ $(14 \div 7) - 1 \times 1^2$

⑨ $(18^2 \div 18) - (3^2 + 8)$

⑩ $35 + (9^2 - 7^2) + 6$

Day 24

Parentheses

Name: ____________________

Score:

① $8^2 - (9^2 \div 3) + 5^2$

② $(11 + 1^2) \div 3 + 6$

③ $32 + (4 \times 3^2 \div 2)$

④ $(42 + 7) - (1^2 \times 7^2)$

⑤ $(4^2 - 14 + 4^2) \div 6$

⑥ $1 \times 5^2 \div (50 - 5^2)$

⑦ $(34 - 6) \div 2^2 - 2$

⑧ $(10^2 \div 50) + (3 \times 10)$

⑨ $36 \div (5^2 - 7 \times 3)$

⑩ $(7 \times 4 - 11 + 12)^2$

Day 25

Parentheses

Name: ________________

Score:

① $35 \div 7 (50 - 7^2)$

② $10(9^2 \div 9^2 - 1)$

③ $(50 + 3^2 \times 2) \div 2^2$

④ $8^2 \div (28 - 26) + 1^2$

⑤ $(18 - 4^2) (5 + 12)$

⑥ $(50 - 6^2 + 1) - 3$

⑦ $10(22 \div 11) + 5^2$

⑧ $18 - (3^2 \times 9 \div 9)$

⑨ $(6^2 \div 18 - 2) + 27$

⑩ $(50 \times 4) - 11 + 9^2$

Day 26
Parentheses

Name: ____________________

Score:

① $7^2 \times 1 - (22 + 9)$

② $16 \div 8(3 \times 2^2)$

③ $28 - (4 \times 4 \div 4)^2$

④ $(6 + 5)^2 - (1 + 6)^2$

⑤ $(9^2 \times 1^2 \div 3^2) + 10$

⑥ $(38 - 28) \times 6^2 \div 36$

⑦ $296 - (21 + 10 \times 5^2)$

⑧ $12 + (6^2 \div 2^2 - 9)$

⑨ $7^2 - (8 + 7) \times 3$

⑩ $(2 \times 7)(18 - 17)$

Day 27

Parentheses

Name: ______________________

Score:

① $(16 - 10)^2 + (13 + 20)$

② $2^2 \div (1 + 1 \times 1)$

③ $5 + (18 \div 3)^2 \times 3$

④ $(16 \times 5^2 - 8) + 70$

⑤ $(11 + 12)^2 - (12 \div 12)^2$

⑥ $37 - (20 \times 4) \div 8$

⑦ $10^2 \div (5 + 3 \times 5)$

⑧ $27 + 7(2 \times 1)$

⑨ $30 + (39 - 9) \div 10$

⑩ $(15 \times 6 - 40) + 7$

Day 28

Parentheses

Name: ____________________

Score:

① $29 + 36 - (10^2 \div 50)$

② $(3^2 - 6) - (1 + 1^2)$

③ $(4 \times 2^2 + 3) - 4^2$

④ $(81 \div 9 + 6^2) \times 1$

⑤ $(7^2 - 49) + (2 \times 1)^2$

⑥ $(2 + 8)(8 + 9)$

⑦ $10^2 \times (5 - 10 \div 5)$

⑧ $(4 + 3 \times 8) \div 7$

⑨ $6^2 + (6^2 - 4 \div 4)$

⑩ $(2^2)(81 \div 3^2) - 20$

Day 29

Parentheses

Name: ____________________

Score:

① $3^2 - (1 \times 2^2 + 1^2)$

② $5 \times 9 + (12 - 2^2)$

③ $(18 + 8 - 19)^2 - 12$

④ $(5 - 1)^2 \div (9 - 1)$

⑤ $11 + (5^2 \div 5^2) - 11$

⑥ $49 \div 7 \times (6^2 - 5^2)$

⑦ $5(2 + 2) \times 3^2$

⑧ $16 + (7^2 \times 1^2 - 9)$

⑨ $(9^2 + 3^2 - 50) \div 8$

⑩ $(9 - 2)^2 - (9 \times 2)$

Name: ___________________

Score:

① $20 \times 6^2 - (8 + 1)^2$

② $160 \div (1 + 3 \times 5)$

③ $9 + (3 \times 3 \div 3)^2$

④ $(35 + 4^2) - (1 + 2^2)^2$

⑤ $(49 - 7^2 + 3) \div 3$

⑥ $(36 \div 6) \times 4 - 2^2$

⑦ $19 + (8^2 - 6 \times 7)$

⑧ $5^2 + (6 \div 3) - 5^2$

⑨ $(5^2 - 24 + 7) \div 4$

⑩ $(1 \times 6)^2 - (4 - 2)^2$

Day 31

Parentheses

Name: ____________________

Score:

① $(349 - 7^2)(1 + 2)$

② $(3 + 2)^2 \div (30 - 5^2)$

③ $(6^2 - 1^2 + 5^2) \div 30$

④ $(9 \div 3^2 + 3^2) \times 4$

⑤ $(10 - 8)^2 + (2 - 1)^2$

⑥ $(6^2 \div 3) - (4 \times 1^2)$

⑦ $45 + (6 \times 2^2 + 2^2)$

⑧ $50 \div (36 - 26)(7)$

⑨ $7 + (43 - 14 \div 2)$

⑩ $100(81 \div 9^2 - 1^2)$

Day 32

Parentheses

Name: ____________________

Score:

① $(6 + 4)^2 - (5 \div 1)^2$

② $28 - (49 \div 7^2 \times 28)$

③ $3 \div (2 - 1)^2 \times 3^2$

④ $(4 \times 2 - 2)^2 + 2$

⑤ $(5 \times 6 - 2) \div (7 - 5)$

⑥ $6 + 12 \div (10 - 7)$

⑦ $(4)(7^2 + 1^2 \times 3)$

⑧ $27 - (77 \div 11)(2)$

⑨ $80 \div (1 + 3) + 10$

⑩ $(7 \times 2 - 2) \div 6$

Day 33

Parentheses

Name: ____________________

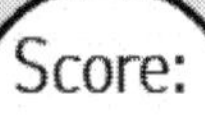

① $18 \div 9 + (4 - 3)^2$

② $3(1 + 7 \times 2)$

③ $6^2 + (4^2 - 2 \times 3)$

④ $(8 + 0)^2 - (24 + 27)$

⑤ $(35 + 25 - 12) \div 8$

⑥ $150 \div 15 \times (4^2 - 2^2)$

⑦ $19 - (3 \div 3 \times 1)^2$

⑧ $5 + (6 \div 6) - 2^2$

⑨ $(3 + 15 \div 3)^2 + 16$

⑩ $(7^2 \times 4^2) + (30 - 21)$

Day 34

Parentheses

Name: ____________________

Score:

① $(8 - 8) \times (3^2 + 5^2)$

② $28 \div (10 - 3 \times 2)$

③ $6 + (9 - 4) \times 7$

④ $(28 - 21 + 1) \times 3^2$

⑤ $(8^2 + 1^2) - (12 \div 3)$

⑥ $12 - (6^2 \div 3) \div 2$

⑦ $30 + (5 - 3 \div 1)^2$

⑧ $2 + 5(2 \times 1)$

⑨ $6^2 + (5^2 - 7) \div 6$

⑩ $(30 - 10 \times 2) + 7$

Day 35
Parentheses

Name: ____________________

Score:

① $12 \div (2 \times 3 - 4)^2$

② $30 - 4^2 \div (14 - 6)$

③ $(9^2 + 23 - 49) \div 5$

④ $(7 + 2)(1 + 4)$

⑤ $8^2 + (81 \div 9) - 7^2$

⑥ $60 \div 3 \times (28 \div 7)$

⑦ $8 \div (6 \div 3)^2 \times 5$

⑧ $16 + (3 + 2^2 \times 9)$

⑨ $(9^2 + 3^2 - 50) \div 2$

⑩ $(6^2 - 24) - (3^2 \times 1)$

Day 36

Nested Parentheses

Name: ____________________

Score:

① $((6 + 7^2) \div 5)2$

② $7((40 - 6^2) \div 2)$

③ $4(72 - (4 \times 7 + 6^2))$

④ $(8^2 - 59)((3 + 3^2) \div 2)$

⑤ $(4 + (6 - 4)) \div 3$

⑥ $2^2((7 - 2)2)$

⑦ $36 \div ((3 + 3)^2 \div 4)$

⑧ $56 \div (2^2(5 - 3))$

⑨ $(61 - (7^2 + 6))10$

⑩ $(12 - 4) \div ((6 + 2) \div 4)$

Day 37

Nested Parentheses

Name: ____________________

Score:

① $((31 + 23)) - (9 \div 3)^2$

② $80 - ((6 + 2)^2 + 2)$

③ $34 - (5^2 + 4^2 \div (1 + 1))$

④ $(3^2 - 2) + (7 \div (4 - 3))^2$

⑤ $((4 \times 2 \div 2)^2 - 4) + 4^2$

⑥ $2((34 - 25) + 9^2 - 8^2)$

⑦ $45 - (3 + (6^2 \div 9)^2)$

⑧ $14 + (6^2 + (8 \div 2 + 33))$

⑨ $150 + ((8 - 2)^2 - 3^2)$

⑩ $(4 + (24 - 6)) \div 2$

Day 38

Nested Parentheses

Name: ____________________ Score:

① $42 - ((8 - 5)^2 + 3)$

② $120 \div ((20 - 2) + 12) - 4$

③ $(16 + (5 + 3^2))2$

④ $36 - ((6 - 3)^2 + (36 - 9))$

⑤ $((7^2 + 10 - 2) + 6) \div 9$

⑥ $3\ ((9 + 1) \times 2^2 - 5^2)$

⑦ $4^2 \div ((2 \times 2) + 2^2)$

⑧ $5((6^2 + 8 \div 2) - 33)$

⑨ $42 \div ((7 - 2)^2 - 19)$

⑩ $(5 + (5 - 1)^2) \div (4 + 3)$

Day 39

Nested Parentheses

Name: ____________________

Score:

① $2(42 + 56 \div (9 - 2))$

② $3((7 + 3) \times 5 - 7^2)$

③ $30 \div ((8 - 5)^2 - 4)$

④ $(100 - (5 + 3)^2) \div 9$

⑤ $10((9 - 3 \div 3) - 1)$

⑥ $9((5 - (1 + 1)^2 - 1))$

⑦ $12 + (3(2 + 1)\ )^2$

⑧ $((2 + 4)^2 + 8) \div (7 - 3)$

⑨ $6 \div (100 \div (7^2 + 1))$

⑩ $3(3^2 + 4(3 + 1))$

Day 40

Nested Parentheses

Name: ____________________

Score:

① $(32 - 5^2)((20 + 28) \div 8)$

② $(1 + 7(13 - 8)) \div 3$

③ $5^2 + (4 \div (3 - 2))^2$

④ $((12 - 4 \div 2) - 2)3$

⑤ $((8 - 5)2)^2 - 32$

⑥ $28 \div ((11 - 9) + (2^2 + 1))$

⑦ $(7^2 - 4) \div ((8 + 7) \div 3)$

⑧ $6(9^2 - (8 \times 7 + 4^2))$

⑨ $9 + (8^2 - 2(10 + 15))$

⑩ $(50 - (5 + 6 \times 2))10$

Day 41

Nested Parentheses

Name: ____________________

Score:

① $30 \div (2 + (8 - 4)^2 \div 2)$

② $3^2 + (3^2 - (8 - 5))$

③ $(26 + 16) \times 1^2 \div (1 \times 1)^2$

④ $((6 \div 2) - 3) + 1^2 - 1^2$

⑤ $(8 - (3 - 2))^2 + 37$

⑥ $120 \div ((20 - 2) + 12) - 1$

⑦ $(5 + 10 - (2 + 6))9$

⑧ $((6 + 6) \div 2)\ (2 + 1)$

⑨ $(4^2 + (38 - 18)) + 4^2$

⑩ $(50 - (45 - 40)) \div (7 - 2)$

Day 42

Nested Parentheses

Name: ____________________

Score:

① $1 - (3^2 \div (14 + 20 - 25))$

② $5((7^2 + 5 \times 3) - 33)$

③ $(55 - 10) + (1^2 \times (23 - 13))$

④ $(8 - (1 + 3))^2 \div 8$

⑤ $(6 - 1) + ((1 + 2) + 26)$

⑥ $4^2 \div ((12 - 5 \times 2) + 6) \times 15$

⑦ $(80 + (4^2 \div 8)) - 6$

⑧ $2(42 + 56 \div (9 - 2))$

⑨ $(6 + (6 - 5))^2 + 8$

⑩ $2((70 - 55) + 5^2 - 10)$

Day 43

Nested Parentheses

Name: ____________________

Score:

① $(5 + (2 \times 14) + (5 \times 3^2))$

② $6^2 \div ((7 - 3)3 - 3)$

③ $(5 \times (48 \div 4))2 - 15$

④ $(56 \div 7)^2 - ((6 \div 6) + 5^2)$

⑤ $100 - 50 + (150 - (50 + 25))$

⑥ $(22 + (8 \times 2^2)) - (4^2 + 16)$

⑦ $(100 - (4^2 + 9 \times 8))2$

⑧ $(3 \times 2)^2 - (45 \div (7 - 2))$

⑨ $(4 + (30 - 12)) \div 2$

⑩ $2((20 - 14) + 7 - 3^2)$

Day 44

Nested Parentheses

Name: ________________

Score:

① $56 \div ((6 - 2)^2 \div 2)$

② $((3 \times 2)^2 - 16) \div 5$

③ $23 + ((27 \div 9) \times 5^2) - 8^2$

④ $(150 \div 50) \times (4 + (27 - 23))$

⑤ $((3^2 - 6) \times (4 + 2^2)) - 24$

⑥ $6^2 \div (16 \div (3 + 1))$

⑦ $(61 + (8 - 6))10$

⑧ $9 \times 3 + (80 \div (16 - 14))$

⑨ $10^2 + ((32 + 68) \div 4)$

⑩ $(72 - (4^2 + 4 \times 7)) \div 2$

Day 45

Nested Parentheses

Name: ____________________

Score:

① $44 - 28 + ((8 \div 2)^2 - 13)$

② $50 - ((12 - 7)^2 + 6)$

③ $(10^2 \div (37 - 17)) + 5 \times 8$

④ $90 + ((3 \div 1 + 6)^2 + 2)$

⑤ $9 \times 6 + (8 - (3 + 4))$

⑥ $48 \div ((18 - 4^2) + (25 - 15))$

⑦ $((20 + 10 - 2) + 8) \div 6$

⑧ $(6 + (12 + 7))5$

⑨ $(7 - 4)\ ((8 + 16) \div 2)$

⑩ $10(2 + (17 - 9))$

Day 46

Nested Parentheses

Name: ____________________

Score:

① $2 + (6 + (3 \times 2)^2) \div 7$

② $(12 - (1 + 7))^2 \div 2$

③ $100 - ((45 + 40) - (4 \times 2)^2)$

④ $(14 - 6) - (1 + (27 - 22))$

⑤ $(7^2 + (24 - 13)) + 3^2$

⑥ $((15 + 20) \div 5)\ (3 + 4)$

⑦ $(3 + 2)^2 - (10 - (9 - 3))^2$

⑧ $(8 - (3 - 2))^2 + 37$

⑨ $8^2 - 1^2 + (24 \div (7 - 1))$

⑩ $90 \div ((38 + 41) - 7^2)$

Day 47

Nested Parentheses

Name: ____________________

Score:

① $(40 - (30 - 8)) \div (7 - 4)$

② $48 \div (1 \times (4^2 - 10))$

③ $(5^2 + (3 + 3))^2 - (2 \times 2)^2$

④ $4 + (43 - (2 \times 3)^2 \div 6^2)$

⑤ $((22 - 17) \times 3) - (18 \div 9)$

⑥ $(1 + 2)\ (5 \times (5 - 4))$

⑦ $4^2 + (5^2 - (20 - 5))$

⑧ $50 - (10 + 2(10 \div 10))$

⑨ $20 + ((3^2 - 2)^2 - (4 \times 2))$

⑩ $(40 - 2(4 + 3^2)) + 34$

Day 48

Nested Parentheses

Name: ____________________

Score:

① $100 - ((3 \times 4 \div 2)^2 + 4^2)$

② $4((5 + 3) \div 2) - 9$

③ $(81 \div (4^2 - (2^2 \times 3) + 5))$

④ $(20 - 12)((3 + 7) \div 2)$

⑤ $22 + ((6 \div 2)^2 - (1 + 5))$

⑥ $(52 - (13 - 6)) \div (8 + 1^2)$

⑦ $3 + (5^2 - (5 + 40 \div 4))$

⑧ $(8 - (3 - 2))^2 + 37$

⑨ $24 \div ((6 - 4 + 2) \times 2)$

⑩ $((1 + 3) \times 2)^2 - 34$

Day 49

Nested Parentheses

Name: ________________

Score:

① $((20 + 70) - (3 + 9 \times 8)) \div 3$

② $(5 + 10 - (2 + 6))9$

③ $(41 - (14 - 5)) \div (12 - 8)$

④ $(9^2 - (5 + 9 \times 7)) + 4^2$

⑤ $(2 \times (6 - 1))^2 \div 4$

⑥ $(18 + 22) \div (4 \times (34 - 24))$

⑦ $4^2 \div ((5 \times 7 + 1) - 20)$

⑧ $((10 - 5) \times ((9 \div 3)^2 + 1))$

⑨ $2((2 + 2 - 2)^2 \div 2)$

⑩ $((50 - 5 \times 8) + 30)2$

Day 50

Nested Parentheses

Name: ________________

Score:

① $(4 \times 5 - (2 + 14))(5 + 2)$

② $7((4^2 + 10 \div 2) - 10)$

③ $120 \div ((6 + 3 \times 3)2)$

④ $37 - (6 \div 3 \times (3 + 1^2)^2)$

⑤ $((7 - 5)^2 + (5 \times 7)) \div 3$

⑥ $(2(42 - 24)) + 8 \div 4$

⑦ $81 \times 2 - ((6 \times 2)^2 + 6)$

⑧ $43 - ((30 - 9) \div 3 + 2)$

⑨ $(9 + 1)^2 - (15 - (7 - 4))$

⑩ $27 + (12 + 30 \div (3 - 2))$

Day 51

5-Step Problems

Name: ____________________

Score:

① $(100 + 50) \div 2 - 3(5 - 1)$

② $(3 + 5) \times 2(40 - 4) \div 3^2$

③ $30 - 6 + 4^2 - 3 \times 7 + 10$

④ $(6^2 - 9 \times 3)\ (8 + 2 \div 2)$

⑤ $(22 + 8)5 \div 25 - 2 \times 2$

⑥ $24 \div (40 - 4^2 \times 2)\ (3 + 7)$

⑦ $(37 + 5) \div (3^2 - 4 \times 4 \div 8)$

⑧ $14 \div 2 + 3 \times 8 - 20 \div 4$

Day 52

5-Step Problems

Name: ________________

Score:

① $5 - (1 \times 2) + 3 \div (2 - 1)$

② $4 \div 1 \times 3^2 \div (2 + 7) - 3$

③ $40 + 27 - (5 \times 3) + (8 \times 1^2)$

④ $(4 \times 2 + 2)^2 \div 20 - 1 + 1$

⑤ $9 - 9 + 1 \times 2\ (4 \times 3)^2$

⑥ $(8 + 7 \times 4 - 2) - 3^2 \div 9$

⑦ $1 \times 6 - 2 + (80 \div 40 + 3)$

⑧ $7 \div (6 - 5) + 28 \times (5 \div 5)^2$

Day 53

5-Step Problems

Name: ____________________

Score:

① $9^2 \times 2 \div 81 - 1 + 6^2 - 30$

② $5 + 7 - 8 \times 3 \div 24 + 2^2$

③ $13 - (42 - 6 \times 7) + 5 \times 5$

④ $18 \div 9 + 5 \times (4 + 4 - 5)^2$

⑤ $37 \div (10 + 11 \times 1 + 16 \times 1)$

⑥ $(2 \times 2)^2 + 5 - 13 \div 13 + 17$

⑦ $29 + (7 + 8 - 1 \times 1)^2 \div 98$

⑧ $(4 - 3)^2 + (9 \div 9)^2 \times (1 \times 3)$

Day 54

5-Step Problems

Name: ____________________

Score:

① $(9 + 2)^2 - 6 \div 1 \times (35 - 18)$

② $8 \times 5 + (4 - 2)^2 - 2 \div 1$

③ $49 \div 7^2 + 3 - 18 \div (4^2 + 2)$

④ $28 - 5 \times 2 \div 10 + 4 \times 5$

⑤ $(23 - 24 \div 8 + 10) \times 2 - 5$

⑥ $34 + 15 - (2 \times 7 - 1)^2 \div 169$

⑦ $4 \times 7 + 4^2 - 195 \div (1^2 + 4)$

⑧ $30 \div 2 - 3 \times 2 + 20 \times 1$

Day 55

5-Step Problems

Name: ____________________

Score:

① $100 \div 10 + (8 \times 4) - 36 \div 6$

② $(150 + 40 \div 8 - 6 + 8) - 157$

③ $3 \times (2 - 1)^2 + (34 \div 17 - 1^2)$

④ $(50 - 9 \times 5) - 64 \div 4^2 + 8^2$

⑤ $8 \div 4 \times 14 - 16 + 1 \times 11$

⑥ $7^2 - (26 + 74) \div 20 \times (7 \div 1)$

⑦ $19 + (8 - 8 \div 8 \times 2 - 4)^2$

⑧ $5 \times (36 \div 9 + 9 - 12) \times 5$

Day 56

5-Step Problems

Name: ________________

Score:

① $(50 - 30) \times (2 + 1)^2 - 81 \div 9$

② $3^2 \div 9 + (5 \times 3 - 15) \times 3$

③ $26 + 2^2 \times 2 - 7^2 \div 49 - 0$

④ $5 \times (17 - 25 \div 5) \times 1 + 5$

⑤ $10 + 10 - (100 \div 10) + (1 \times 3)^2$

⑥ $81 \div 3^2 + 25 \div 5^2 \times 18 \div 3$

⑦ $(6^2 \div 3) \times (2^2 + 2) - (4 + 3^2)$

⑧ $(15 - 7) \div 8 \times 9 - 9 + 1^2$

Day 57

5-Step Problems

Name: ________________

Score:

① $4 \times 4^2 - 64 + 9^2 \div 3^2 - 4$

② $22 + 23 \times 10^2 \div 100 - 10 + 17$

③ $(5 - 5)^2 + (3 \times 3)^2 \div (3 \times 3)$

④ $(36 \div 6 + 8 \times 1) \times (6 \div 6)$

⑤ $(7 + 5)^2 - (8 \div 2 + 2 \times 8)$

⑥ $9 \div (49 - 40) + (3 + 7 \times 2)^2$

⑦ $29 - 17 + 3 \times 32 \div 96 - 8$

⑧ $(4 \times 2 - 2)^2 - (6 - 3)^2 + 36$

Day 58

5-Step Problems

Name: ____________________

Score:

P.
E.
M.
D
A
S

① $(5^2 - 5) \times 2^2 \div 5 + (7^2 \div 7)$

② $(3^2 \times 4) + (5^2 - 3^2 + 23) - 0$

(9x4) + (25 - 9 + 23) - 0

36 + (25 - 32) - 0

36 + (25 - 32) - 0

36 + (-7) - 0

29 - 0 = 29

LOL it's 73

③ $25 \div 5 + 60 - 5 \times 2 + 19$

④ $36 + (7 - 3)^2 \times (10 \div 2) \times 1$

⑤ $48 - 6 \times (1 \div 1 + 7 \times 1)$

⑥ $3^2 \times 81 \div 3^2 - 26 + 12 - 1^2$

⑦ $6 + (18 - 36 \div 6) + 150 - 50$

⑧ $30 \div 15 + 7 + 5 \times 3^2 \div 3$

Day 59

5-Step Problems

Name: ____________________

Score:

① $2^2 + (26 - 10) \times 2^2 \div (49 - 41)$

② $45 - 5 \times 3^2 + 3^2 - 5^2 \div 5$

③ $4 \times 8 - (18 \div 6) + 5 - 5$

④ $(1 \div 1 \times 7)^2 + 13 - 3 + 6$

⑤ $6^2 + 11 - (6 - 6^2 \div 36) \times 3^2$

⑥ $15 - (6 \times 2 + 3) \div (5 - 2)$

⑦ $(3 \div 3) + (4 + 3 \times 2)^2 + 2$

⑧ $(5 - 1) \times 2 + 2 \div (6 - 4)$

Day 60

5-Step Problems

Name: ____________________

Score:

① $(35 - 17) \times 1 + (50 + 9 \div 3)$

② $(6 + 4 - 2)^2 \div (4^2 + 4^2 - 30)$

③ $49 \div 7 + 11 - 100 \div 10 \times 1$

④ $8 \times 64 \div 4 - 2 \times 8 - 24$

⑤ $3^2 \times (2 - 1 + 2) \times (2 \times 5)$

⑥ $(20 + 10) \div 5 \times 2 \div (2 - 1)^2$

⑦ $32 - 9 \div 3 + 21 \div 7 - 7$

⑧ $2 \times 5 - 2^2 + 10 - 4 \times 2^2$

Day 61

Negative Numbers

Name: ____________________

Score:

① 10 + (5 - (-8))

② ((8 ÷ (-4)) + (-24))

③ ((-3) + (-9 + 3))

④ (-7 x (-36 ÷ -6))

⑤ ((12 - (-3)) + 6)

⑥ ((-15 + -7) - 8)

⑦ (((-7) + 2) x 9)

⑧ (20 ÷ -5 + (-35))

⑨ (2 x (-24 ÷ 8))

⑩ ((-6) + (-10) + 8)

Day 62

Negative Numbers

Name: ____________________

Score:

① 3 - ((-5) + 9)

② ((-3) – 5) + 6

③ ((-7) + (-4)) - 2

④ (-3) + ((-8) + 12)

⑤ (8 - (-7)) + 5

⑥ ((-6) + 4) + (-1)

⑦ ((-3) + 2) + 9

⑧ (5 - (-4)) + 7

⑨ ((-10) – 4) + (-8)

⑩ 3 + ((-5) + 9)

Day 63

Negative Numbers

Name: ____________________

Score:

① -20 - (-15 + 10)

② (35 - 11) - 23

③ (7) + (-4 + 2)

④ 18 - (41 - (-31))

⑤ ((-16) + (-5)) - 6

⑥ ((-25) - 13 - (-12))

⑦ (17 + (-7)) - 3

⑧ ((-28) + (-18) - 8)

⑨ (3 + (-3) - (-2))

⑩ (24 - (-15) - 13)

Day 64

Negative Numbers

Name: ____________________

Score:

① (-36 + (-16) - (-6))

② (5) + (-8 + (-3))

③ 12 - ((-9) - (-7))

④ 47 + ((-13) + 4)

⑤ (1 + (-1) - (-2))

⑥ (6 - 7 + (-9))

⑦ -29 + (14 - (-8))

⑧ (9 - (-3)) - 22

⑨ ((-20) + 10 + (-40))

⑩ ((-3) + (-9 + 3))

Name: ____________________

Score:

① ((8 - 11) + (-6))

② 4 - ((-9) - 7)

③ ((-6) + (-10) + 8)

④ (5 x ((-81) ÷ 9))

⑤ ((8 ÷ (-4)) + (-24))

⑥ (-11 + 21) x (-3)

⑦ 5 x ((-49) ÷ (-7))

⑧ ((-12) ÷ 6) + 39

⑨ ((10 ÷ (-2)) x 10)

⑩ (12- 4 x (-4))

Day 66
Negative Numbers

Name: ____________________

Score:

① ((9 x (-3)) ÷ (-27))

② (4 ÷ (-2) - 36)

③ 18 + ((-2) x (-5))

④ ((7 - (-9) + 6))

⑤ (-6 x 8 ÷ 12)

⑥ 11 - (21 + (-3))

⑦ (9 ÷ 3 - (-31))

⑧ ((-16) ÷ 8) x 2

⑨ ((-26) + 14 - (-14))

⑩ (3 +(-6) ÷ (-3))

Day 67

Negative Numbers

Name: ________________

Score:

① ((24 - 44) x (-4))

② (2 x (-24 ÷ 8))

③ (20 - 40 + (-25))

④ ((-81) ÷ (3 - 2))

⑤ ((-7) + (-7) - 7)

⑥ (8 ÷ (-2)) x 2

⑦ (5 x 5 + (-5))

⑧ (3 x (-36 ÷ 6))

⑨ ((-42÷ 6) + 15)

⑩ (13 -((-4) - 37))

Day 68

Negative Numbers

Name: ______________

Score:

① ((-18 ÷ 9) - 7)

② (5 x 2 ÷ -10)

③ (4 + (-3) x -6)

④ (20 ÷ -5 + (-35))

⑤ ((8 - 2) ÷ -6)

⑥ (-7 x (-36 ÷ -6))

⑦ (100 ÷ -50 - 10)

⑧ ((17 + (-19)) x 2)

⑨ ((3 + 9) - (-39))

⑩ ((-34) + (-13) + 7)

Day 69

Negative Numbers

Name: ____________________

Score:

① (-1 x 5 x 4)

② ((-5) - ((-2) + (-10)))

③ ((24 ÷ 6) + (-24))

④ ((-0) - ((-50) ÷ -5))

⑤ ((-15 + -7) - 8)

⑥ (37 + 48 + (-17))

⑦ ((1) ((-14) + (-42)))

⑧ (10 ÷ ((-6) - 4))

⑨ ((-11) - 6) ÷ -17

⑩ (7 - (-3) + (-2))

Day 70

Negative Numbers

Name: ____________________

Score:

① $(((-19) - (-17)) + (-15))$

② $((8 \times -4) \div -16)$

③ $(3 - 5 \times (-6))$

④ $(((-7) + 2) \times 9)$

⑤ $((-1) + ((-6) - 7))$

⑥ $(-40 \div 4 \times -3)$

⑦ $((12 - (-3)) + 6)$

⑧ $((9 \times -2) + (-8))$

⑨ $((-0) + ((-29) \div 1))$

⑩ $((21 \div (7)) - (-21))$

Day 71

Fraction Bars

Name: ________________

Score:

① $\frac{(35 + 45 \div 3)}{5(8 - 6)}$

② $\frac{9 - 32 \div 8 - 2}{5^2 - 3 \times 7}$

③ $\frac{7^2 - 6 \times 8}{6 + 35 \div 7}$

④ $\frac{2(40 - 4)}{3^2}$

⑤ $\frac{2(17 - 6)}{60 \div 5 + 5 \times 3}$

⑥ $\frac{100 - 5^2 \times 3}{4 + 12 \times 2}$

⑦ $\frac{2^2 \times 2}{2(6^2 - 4(5 + 1))}$

⑧ $\frac{11 + 3 \times 9 - 2}{(53 - 8) \div 5}$

Day 72

Fraction Bars

Name: ____________________

Score:

① $\dfrac{3^2 + 13}{50 - 5^2}$

② $\dfrac{(22 \div 11 \times 7) + 4}{44 - (6 \times 7)}$

③ $\dfrac{8^2 \div (7 + 9)}{5 \times 3 - 2^2}$

④ $\dfrac{11 - 6 + 6}{28}$

⑤ $\dfrac{(13 - 3) + 3^2}{4 \times 7 + 8 - 15}$

⑥ $\dfrac{(8^2 + (45 \div 15))}{(39 - 29) \times 7}$

⑦ $\dfrac{(2 \times 5) - (8 \div 4)}{49 \div 7^2 + 7}$

⑧ $\dfrac{24 - 4^2 \div 4 + 30}{15 + (9 \times 4)}$

Day 73

Fraction Bars

Name: ____________________

Score:

① $\dfrac{(20 \times 2) - 35}{9 + 7 \div 7}$

② $\dfrac{6 + (8 - 4) \times 6}{10 \div 5 + 8}$

③ $\dfrac{36 - 10 + 8^2}{15 \times 3^2 \div 9}$

④ $\dfrac{4^2 \div 2 \times 5 - 7}{(28 - 25 + 7)^2}$

⑤ $\dfrac{9 \times (4 - 2)^2}{(30 \div 15 - 2) + 36}$

⑥ $\dfrac{(16 - 6)^2 \times 2}{5 \times (24 - 14)}$

⑦ $\dfrac{(7 + (7 \times 3) \div 21)}{((91 - 10) \div 3^2)}$

⑧ $\dfrac{(30 \div 6 + 2)^2 - 5^2}{3 \times 9 - 1}$

Day 74

Fraction Bars

Name: ________________

Score:

① $\frac{((12 \div 6) \times 3)}{90 \div (1 \times 10)}$

② $\frac{25 - 13 + (16 \div 8)}{(15 - 9) \times 2^2}$

③ $\frac{(4 \times 1)^2 \div 2}{5^2 - (8 + 9)}$

④ $\frac{18 - 7^2 \div 7 + 3}{8 \times 9 \div 2}$

⑤ $\frac{(8 \div 2)^2 - 6}{(8 + 7)2 - 25}$

⑥ $\frac{(3^2 + 16 - 3^2)}{39 - 10 + 4}$

⑦ $\frac{19 \times 3 - 49 \div 7}{9 \times 9 \div 81}$

⑧ $\frac{12 - 10 + 1^2 \times 4}{(11 - 7)^2 \times 2}$

Day 75

Fraction Bars

Name: ____________________

Score:

① $\dfrac{18 + (5 \times 2)^2}{59}$

② $\dfrac{36 \div 12 - 1}{(3 + 2)13}$

③ $\dfrac{(7^2 + 11)4}{(12 - 10)^2}$

④ $\dfrac{(4 + 5)^2}{(9 - 2) \times (8 + 5)}$

⑤ $\dfrac{6 \times 3 \div 9}{28 \div (4 - 2)^2 + 2}$

⑥ $\dfrac{(18 - (8 + 6))}{49 \div 7}$

⑦ $\dfrac{8 + ((5 \times 5) - 8)}{11 - (7 + 3)}$

⑧ $\dfrac{(9^2 - (4 + 2 \times 2)^2)}{5((7 - 3)^2 + 2^2)}$

Day 76

Fraction Bars

Name: ______________ Score:

① $\frac{(8 - 6 \div 3)^2}{(2 + 2)^2 \div (2 \times 2)}$

② $\frac{((4^2 \times 2) + 7)}{(6^2 + 3)}$

③ $\frac{4 \times 5 - 6}{17 + 6 - 9 \div 3}$

④ $\frac{12 \div 3 + 26}{(2 \times 5)^2 \div 20}$

⑤ $\frac{150 - (7 + 4)^2 \times 1}{6^2}$

⑥ $\frac{6\,(4 - 3)}{3 \div 3(3)}$

⑦ $\frac{8^2}{8^2 \times 6 \div 96}$

⑧ $\frac{22 \div 11 \times 6 - 10}{0 + 11}$

Day 77

Fraction Bars

Name: ________________ Score:

① $\dfrac{6^2 - 7 \times 5}{14 \div 7 \times 4^2}$

② $\dfrac{8\,(5) \div 10}{5^2}$

③ $\dfrac{((23 + 19) - 34)}{2 \times (16 \div 8)^2}$

④ $\dfrac{(24 \times 3) \div 3^2}{96 \div (2^2 \times 6)}$

⑤ $\dfrac{(21 \div (2 + 5))}{41 + 80 \div 10}$

⑥ $\dfrac{(5)1^2}{((25 - 13) + 7)}$

⑦ $\dfrac{8 \times 2^2 - 7^2 \div 49 + 1}{5^2 - 17}$

⑧ $\dfrac{(28 - 16) \times 8 - 15}{(6 \div 6 + 8)^2}$

Day 78

Fraction Bars

Name: ____________________

① $\dfrac{21 + (9 \div 3)}{(16 - 13)4}$

② $\dfrac{(6 \div 3 + 2 - 2)^2}{(5 - 4 \times 1)^2}$

③ $\dfrac{38 + 23 - 7 \times 8}{(8)\,(5)}$

④ $\dfrac{20^2 \div 100}{42 + 8^2 - 24}$

⑤ $\dfrac{2^2 \times (4^2 + (8 \div 2))}{((9 \div 3) + 3^2) - 2^2}$

⑥ $\dfrac{39 + 7^2 - 49}{9^2 \div 3 + 27}$

⑦ $\dfrac{8^2 \times 5}{32}$

⑧ $\dfrac{(12 - 6) + (12 - 6)}{15 \times 2 + 5}$

Day 79

Fraction Bars

Name: ______________________

Score:

① $\dfrac{108 \div 6^2 - 1}{3^2 - 5 + 16}$

② $\dfrac{3 \times 5 + 2}{((2 + 2)^2 - 8) \times 3}$

③ $\dfrac{(4^2 + 11) \div 3^2}{8 \times 6^2 - 9^2}$

④ $\dfrac{38 - 33 + 4}{72 \div 8}$

⑤ $\dfrac{((3 \times 8) - 15)}{(24 - (10 \div 5) + 11)}$

⑥ $\dfrac{20 \div 4 + 20}{(15 - 10)^2 \times 5}$

⑦ $\dfrac{(23 - 13) \times 6 \div 2}{5 \times 2}$

⑧ $\dfrac{1(13 - 3 + 5)}{(18 \div 2)3}$

Day 80

Fraction Bars

Name: ____________________

Score:

① $\dfrac{9 - 2 \times 4}{14 \div 7 \times 4^2}$

② $\dfrac{(3 + 3) \div 1 - 2}{5^2 \div 5 + 3}$

③ $\dfrac{10 + (9 - 5)}{21 \times 3}$

④ $\dfrac{19 + 13 - 11}{3^2 \times (6 + 4)}$

⑤ $\dfrac{4^2 + 7^2}{74 + 8 \div 2 - 5}$

⑥ $\dfrac{(5 + 1) \times (11 - 9)}{10 - 3 + 8}$

⑦ $\dfrac{8 \times 2^2 - 14}{3^2}$

⑧ $\dfrac{3 + (4 \times 4) \div 2}{28 \div 7 + 10}$

Day 1

① 13 + 7 x 3
13 + 21
34

② 9 × 2 - 13
18 - 13
5

③ 32 ÷ 8 + 4
4 + 4
8

④ 29 - 24 ÷ 2
29 - 12
17

⑤ 20 ÷ 4 x 9
5 x 9
45

⑥ 15 - 7 x 2
15 - 14
1

⑦ 14 + 8 ÷ 2
14 + 4
18

⑧ 45 ÷ 9 x 2
5 x 2
10

⑨ 30 - 20 ÷ 5
30 - 4
26

⑩ 24 + 6 x 6
24 + 36
60

Day 2

① 45 - 40 ÷ 8
45 - 5
40

② 28 + 8 ÷ 4
28 + 2
30

③ 3 + 2 x 9
3 + 18
21

④ 6 x 4 - 7
24 - 7
17

⑤ 10 ÷ 2 x 5
5 x 5
25

⑥ 10 - 2 x 4
10 - 8
2

⑦ 16 + 4 ÷ 2
16 + 2
18

⑧ 24 ÷ 3 x 2
8 x 2
16

⑨ 12 - 4 ÷ 4
12 - 1
11

⑩ 2 + 2 x 3
2 + 6
8

Day 3

① 5 + 5 x 6
5 + 30
35

② 3 x 5 - 4
15 - 4
11

③ 49 ÷ 7 + 10
7 + 10
17

④ 19 - 21 ÷ 3
19 - 7
12

⑤ 15 ÷ 15 x 5
1 x 5
5

⑥ 17 - 18 ÷ 3
17 - 6
11

⑦ 33 + 1 x 1
33 + 1
34

⑧ 9 ÷ 3 x 6
3 x 6
18

⑨ 50 - 35 x 1
50 - 35
15

⑩ 13 + 12 ÷ 4
13 + 3
16

Day 4

① 28 - 16 ÷ 8
28 - 2
26

② 16 x 4 + 20
64 + 20
84

③ 4 + 1 - 2
5 - 2
3

④ 18 ÷ 9 x 2
2 x 2
4

⑤ 28 - 3 x 9
28 - 27
1

⑥ 2 + 7 x 18
2 + 126
128

⑦ 9 x 3 + 5
27 + 5
32

⑧ 16 ÷ 8 - 2
2 - 2
0

⑨ 20 - 15 ÷ 3
20 - 5
15

⑩ 40 + 2 - 28
42 - 28
14

Day 5

① 30 x 4 + 5
120 + 5
125

② 5 - 3 ÷ 3
5 - 1
4

③ 8 + 1 x 10
8 + 10
18

④ 16 ÷ 8 + 27
2 + 27
29

⑤ 50 x 3 ÷ 15
150 ÷ 15
10

⑥ 44 - 40 ÷ 4
44 - 10
34

⑦ 17 + 20 - 7
37 - 7
30

⑧ 16 x 8 - 50
128 - 50
78

⑨ 5 - 2 x 1
5 - 2
3

⑩ 6 + 20 - 20
26 - 20
6

Day 6

① 6 + 10 - 18 ÷ 3
6 + 10 - 6
16 - 6
10

② 13 + 7 x 15 ÷ 5
13 + 105 ÷ 5
13 + 21
34

③ 28 - 12 x 2 + 6
28 - 24 + 6
4 + 6
10

④ 32 ÷ 8 x 5 + 4
4 x 5 + 4
20 + 4
24

⑤ 26 + 8 x 24 ÷ 6
26 + 192 ÷ 6
26 + 32
58

⑥ 20 ÷ 4 x 11 + 3
5 x 11 + 3
55 + 3
58

⑦ 4 + 8 ÷ 2 x 3
4 + 4 x 3
4 + 12
16

⑧ 14 + 14 ÷ 7 - 2
14 + 2 - 2
16 - 2
14

⑨ 3 x 12 + 4 x 10
36 + 4 x 10
36 + 40
76

⑩ 10 x 2 - 16 ÷ 4
20 - 16 ÷ 4
20 - 4
16

Day 7

① 3 x 5 - 15 ÷ 5
15 - 15 ÷ 5
15 - 3
12

② 48 - 40 + 2 x 4
48 - 40 + 8
8 + 8
16

③ 17 - 6 ÷ 3 x 1
17 - 2 x 1
17 - 2
15

④ 5 + 37 x 24 - 23
5 + 888 - 23
893 - 23
870

⑤ 22 - 3 x 5 + 20
22 - 15 + 20
7 + 20
27

⑥ 3 + 9 ÷ 3 x 3
3 + 3 x 3
3 + 9
12

⑦ 9 x 6 - 38 + 5
54 - 38 + 5
16 + 5
21

⑧ 6 - 3 + 8 x 2
6 - 3 + 16
3 + 16
19

⑨ 7 x 5 - 3 ÷ 3
35 - 3 ÷ 3
35 - 1
34

⑩ 1 + 49 x 4 ÷ 4
1 + 196 ÷ 4
1 + 49
50

Day 8

① 28 + 6 x 3 - 1
28 + 18 - 1
46 - 1
45

② 9 x 6 ÷ 2 + 4
54 ÷ 2 + 4
27 + 4
31

③ 36 ÷ 6 + 15 x 2
6 + 15 x 2
6 + 30
36

④ 11 - 5 x 2 + 3
11 - 10 + 3
1 + 3
4

⑤ 22 + 18 - 40 ÷ 8
22 + 18 - 5
40 - 5
35

⑥ 35 - 5 ÷ 1 x 7
35 - 5 x 7
35 - 35
0

⑦ 50 ÷ 50 x 20 + 20
1 x 20 + 20
20 + 20
40

⑧ 7 x 16 + 21 - 11
112 + 21 - 11
133 - 11
122

⑨ 9 + 2 - 1 x 8
9 + 2 - 8
11 - 8
3

⑩ 13 - 5 x 0 + 34
13 - 0 + 34
13 + 34
47

Day 9

① 7 x 13 - 14 ÷ 7
91 - 14 ÷ 7
91 - 2
89

② 27 + 0 x 6 - 6
27 + 0 - 6
27 - 6
21

③ 41 - 20 ÷ 10 x 9
41 - 2 x 9
41 - 18
23

④ 8 ÷ 8 + 3 x 3
1 + 3 x 3
1 + 9
10

⑤ 2 x 22 + 17 - 16
44 + 17 - 16
61 - 16
45

⑥ 5 + 39 - 36 ÷ 9
5 + 39 - 4
44 - 4
40

⑦ 44 - 2 ÷ 2 + 4
44 - 1 + 4
43 + 4
47

⑧ 12 ÷ 6 x 37 - 33
2 x 37 - 33
74 - 33
41

⑨ 5 + 3 - 15 ÷ 5
5 + 3 - 3
8 - 3
5

⑩ 6 x 10 + 23 - 5
60 + 23 - 5
83 - 5
78

Day 10

① 36 ÷ 9 + 11 x 3
4 + 11 x 3
4 + 33
37

② 18 - 2 x 6 + 48
18 - 12 + 48
6 + 48
54

③ 12 x 1 - 6 + 2
12 - 6 + 2
6 + 2
8

④ 4 + 14 ÷ 1 x 8
4 + 14 x 8
4 + 112
116

⑤ 9 - 3 x 3 + 6
9 - 9 + 6
0 + 6
6

⑥ 36 ÷ 12 + 24 - 11
3 + 24 - 11
27 - 11
16

⑦ 17 x 3 - 19 + 25
51 - 19 + 25
32 + 25
57

⑧ 16 + 16 ÷ 16 - 8
16 + 1 - 8
17 - 8
9

⑨ 41 - 12 + 1 x 0
41 - 12 + 0
29 + 0
29

⑩ 2 x 10 - 18 ÷ 2
20 - 18 ÷ 2
20 - 9
11

Day 11

① 1 + 17 x 12 - 8
1 + 204 - 8
205 - 8
197

② 35 ÷ 5 - 2 x 3
7 - 2 x 3
7 - 6
1

③ 6 x 1 + 6 ÷ 2
6 + 6 ÷ 2
6 + 3
9

④ 26 - 12 ÷ 6 + 4
26 - 2 + 4
24 + 4
28

⑤ 43 + 8 - 16 x 3
43 + 8 - 48
51 - 48
3

⑥ 9 ÷ 9 x 2 - 2
1 x 2 - 2
2 - 2
0

⑦ 1 x 23 + 6 - 1
23 + 6 - 1
29 - 1
28

⑧ 14 - 3 x 2 + 17
14 - 6 + 17
8 + 17
25

⑨ 50 ÷ 25 - 1 x 1
2 - 1 x 1
2 – 1
1

⑩ 13 - 15 ÷ 3 + 5
13 - 5 + 5
8 + 5
13

Day 12

① 7 x 7 - 34 ÷ 2
49 - 34 ÷ 2
49 - 17
32

② 24 - 4 ÷ 2 x 1
24 - 2 x 1
24 - 2
22

③ 3 + 9 x 9 - 3
3 + 81 - 3
84 - 3
81

④ 45 ÷ 5 + 0 - 0
9 + 0 - 0
9 - 0
9

⑤ 20 - 4 ÷ 2 x 8
20 - 2 x 8
20 - 16
4

⑥ 11 x 6 - 26 + 6
66 - 26 + 6
40 + 6
46

⑦ 8 + 19 x 21 ÷ 7
8 + 399 ÷ 7
8 + 57
65

⑧ 14 ÷ 14 + 3 - 2
1 + 3 - 2
4 - 2
2

⑨ 8 - 4 ÷ 1 + 1
8 - 4 + 1
4 + 1
5

⑩ 26 x 2 - 37 + 8
52 - 37 + 8
15 + 8
23

Day 13

① 50 - 26 + 8 ÷ 4
50 - 26 + 2
24 + 2
26

② 28 + 18 ÷ 9 - 28
28 + 2 - 28
30 - 28
2

③ 36 ÷ 6 - 2 + 1
6 - 2 + 1
4 + 1
5

④ 1 x 9 + 13 - 15
9 + 13 - 15
22 - 15
7

⑤ 13 - 8 x 1 + 2
13 - 8 + 2
5 + 2
7

⑥ 41 + 6 - 8 x 3
41 + 6 - 24
47 - 24
23

⑦ 1 x 25 ÷ 5 + 1
25 ÷ 5 + 1
5 + 1
6

⑧ 33 ÷ 3 + 27 x 3
11 + 27 x 3
11 + 81
92

⑨ 41 + 16 - 16 ÷ 4
41 + 16 - 4
57 – 4
53

⑩ 19 - 7 x 2 + 7
19 - 14 + 7
5 + 7
12

Day 14

① 6 x 9 - 50 + 2
54 - 50 + 2
4 + 2
6

② 19 - 18 x 3 ÷ 3
19 - 54 ÷ 3
19 - 18
1

③ 42 + 49 ÷ 7 - 40
42 + 7 - 40
49 - 40
9

④ 38 ÷ 2 - 1 x 5
19 - 1 x 5
19 - 5
14

⑤ 8 x 7 + 16 ÷ 8
56 + 16 ÷ 8
56 + 2
58

⑥ 15 - 14 x 1 + 4
15 - 14 + 4
1 + 4
5

⑦ 10 - 4 ÷ 1 x 1
10 - 4 x 1
10 - 4
6

⑧ 17 + 6 - 30 ÷ 5
17 + 6 - 6
23 - 6
17

⑨ 11 x 2 + 25 - 22
22 + 25 - 22
47 - 22
25

⑩ 44 ÷ 4 x 13 + 7
11 x 13 + 7
143 + 7
150

Day 15

① 35 - 14 ÷ 2 x 5
35 - 7 x 5
35 - 35
0

② 12 + 37 x 24 ÷ 8
12 + 888 ÷ 8
12 + 111
123

③ 4 x 26 - 50 + 40
104 - 50 + 40
54 + 40
94

④ 9 ÷ 9 + 5 - 2
1 + 5 - 2
6 - 2
4

⑤ 0 + 1 x 1 ÷ 1
0 + 1 ÷ 1
0 + 1
1

⑥ 6 x 9 ÷ 9 + 25
54 ÷ 9 + 25
6 + 25
31

⑦ 6 x 6 - 13 + 8
36 - 13 + 8
23 + 8
31

⑧ 27 ÷ 3 + 24 x 4
9 + 24 x 4
9 + 96
105

⑨ 48 - 28 + 5 x 43
48 - 28 + 215
20 + 215
235

⑩ 15 + 5 x 7 - 7
15 + 35 - 7
50 – 7
43

Day 16

① $6 + 7^2 - 18 \div 3$
6 + 49 - 18 ÷ 3
6 + 49 - 6
55 - 6
49

② $24 + 3^2 \times 4 \div 2$
24 + 9 x 4 ÷ 2
24 + 36 ÷ 2
24 + 18
42

③ $21 - 6 \times 3 + 8^2$
21 - 6 x 3 + 64
21 - 18 + 64
3 + 64
67

④ $6^2 \div 3 \times 2 + 2^2$
$36 \div 3 \times 2 + 2^2$
36 ÷ 3 x 2 + 4
12 x 2 + 4
24 + 4
28

⑤ $31 + 4^2 \div 2 \times 6$
31 + 16 ÷ 2 x 6
31 + 8 x 6
31 + 48
79

⑥ $20 \div 4 \times 3^2 - 5^2$
$20 \div 4 \times 9 - 5^2$
20 ÷ 4 x 9 - 25
5 x 9 - 25
45 - 25
20

⑦ $230 - 8 \div 2^2 \times 7$
230 - 8 ÷ 4 x 7
230 - 2 x 7
230 - 14
216

⑧ $9^2 + 24 \div 8 - 2$
81 + 24 ÷ 8 - 2
81 + 3 - 2
84 - 2
82

⑨ $3 \times 5 + 4^2 \times 10$
3 x 5 + 16 x 10
15 + 16 x 10
15 + 160
175

⑩ $5^2 \times 2 - 40 \div 5$
25 x 2 - 40 ÷ 5
50 - 40 ÷ 5
50 - 8
42

Day 17

1. $40 \div 10 + 1 \times 4^2$
   $40 \div 10 + 1 \times 16$
   $4 + 1 \times 16$
   $4 + 16$
   20
2. $5^2 \times 4 \div 20 + 50$
   $25 \times 4 \div 20 + 50$
   $100 \div 20 + 50$
   $5 + 50$
   55
3. $31 + 6 - 9^2 \div 3$
   $31 + 6 - 81 \div 3$
   $31 + 6 - 27$
   $37 - 27$
   10
4. $49 - 2^2 \times 8 + 2$
   $49 - 4 \times 8 + 2$
   $49 - 32 + 2$
   $17 + 2$
   19
5. $1 \times 2^2 + 3^2 - 2$
   $1 \times 4 + 3^2 - 2$
   $1 \times 4 + 9 - 2$
   $4 + 9 - 2$
   $13 - 2$
   11
6. $5^2 + 16 \div 16 - 10$
   $25 + 16 \div 16 - 10$
   $25 + 1 - 10$
   $26 - 10$
   16
7. $18 \div 9 - 1^2 + 1$
   $18 \div 9 - 1 + 1$
   $2 - 1 + 1$
   $1 + 1$
   2
8. $2^2 \times 2^2 - 6 + 5$
   $4 \times 2^2 - 6 + 5$
   $4 \times 4 - 6 + 5$
   $16 - 6 + 5$
   $10 + 5$
   15
9. $7 + 6 \times 6 \div 3^2$
   $7 + 6 \times 6 \div 9$
   $7 + 36 \div 9$
   $7 + 4$
   11
10. $32 - 12 + 6^2 \times 1$
    $32 - 12 + 36 \times 1$
    $32 - 12 + 36$
    $20 + 36$
    56

Day 18

1. $8^2 - 9 \times 7 + 8^2$
   $64 - 9 \times 7 + 8^2$
   $64 - 9 \times 7 + 64$
   $64 - 63 + 64$
   $1 + 64$
   65
2. $2^2 + 3^2 - 18 \div 6$
   $4 + 3^2 - 18 \div 6$
   $4 + 9 - 18 \div 6$
   $4 + 9 - 3$
   $13 - 3$
   10
3. $2 \times 5^2 + 41 - 34$
   $2 \times 25 + 41 - 34$
   $50 + 41 - 34$
   $91 - 34$
   57
4. $16 \div 4^2 + 1 \times 7^2$
   $16 \div 16 + 1 \times 7^2$
   $16 \div 16 + 1 \times 49$
   $1 + 1 \times 49$
   $1 + 49$
   50
5. $14 + 39 - 6^2 \div 3$
   $14 + 39 - 36 \div 3$
   $14 + 39 - 12$
   $53 - 12$
   41
6. $48 - 18 \div 3^2 + 1$
   $48 - 18 \div 9 + 1$
   $48 - 2 + 1$
   $46 + 1$
   47
7. $2 \times 2 + 20 - 4^2$
   $2 \times 2 + 20 - 16$
   $4 + 20 - 16$
   $24 - 16$
   8
8. $27 \div 3^2 - 1 \times 2$
   $27 \div 9 - 1 \times 2$
   $3 - 1 \times 2$
   $3 - 2$
   1
9. $25 - 10 \div 2 + 1^2$
   $25 - 10 \div 2 + 1$
   $25 - 5 + 1$
   $20 + 1$
   21
10. $1^2 + 5 \times 2^2 - 1$
    $1 + 5 \times 2^2 - 1$
    $1 + 5 \times 4 - 1$
    $1 + 20 - 1$
    $21 - 1$
    20

Day 19

1. $26 + 7^2 - 37 \times 1^2$
   $26 + 49 - 37 \times 1^2$
   $26 + 49 - 37 \times 1$
   $26 + 49 - 37$
   $75 - 37$
   38
2. $85 - 15 \times 2^2 + 2$
   $85 - 15 \times 4 + 2$
   $85 - 60 + 2$
   $25 + 2$
   27
3. $49 \div 7^2 + 6^2 - 22$
   $49 \div 49 + 6^2 - 22$
   $49 \div 49 + 36 - 22$
   $1 + 36 - 22$
   $37 - 22$
   15
4. $5^2 \times 3 \div 15 - 5$
   $25 \times 3 \div 15 - 5$
   $75 \div 15 - 5$
   $5 - 5$
   0
5. $20 - 4 \times 5 + 4^2$
   $20 - 4 \times 5 + 16$
   $20 - 20 + 16$
   $0 + 16$
   16
6. $11 + 2^2 - 35 \div 7$
   $11 + 4 - 35 \div 7$
   $11 + 4 - 5$
   $15 - 5$
   10
7. $6^2 \times 1 \div 2 - 4^2$
   $36 \times 1 \div 2 - 4^2$
   $36 \times 1 \div 2 - 16$
   $36 \div 2 - 16$
   $18 - 16$
   2
8. $54 - 27 + 3^2 \times 3$
   $54 - 27 + 9 \times 3$
   $54 - 27 + 27$
   $27 + 27$
   54
9. $14 + 1^2 - 18 \div 3^2$
   $14 + 1 - 18 \div 3^2$
   $14 + 1 - 18 \div 9$
   $14 + 1 - 2$
   $15 - 2$
   13
10. $36 \div 12 \times 5^2 - 50$
    $36 \div 12 \times 25 - 50$
    $3 \times 25 - 50$
    $75 - 50$
    25

Day 20

1. $50 - 7^2 \times 1 + 4$
   $50 - 49 \times 1 + 4$
   $50 - 49 + 4$
   $1 + 4$
   5
2. $2^2 \times 7 + 5^2 - 20$
   $4 \times 7 + 5^2 - 20$
   $4 \times 7 + 25 - 20$
   $28 + 25 - 20$
   $53 - 20$
   33
3. $27 + 2 - 7^2 \div 7$
   $27 + 2 - 49 \div 7$
   $27 + 2 - 7$
   $29 - 7$
   22
4. $10^2 \div 10 \times 1 + 5^2$
   $100 \div 10 \times 1 + 25$
   $100 \div 10 \times 1 + 25$
   $10 \times 1 + 25$
   $10 + 25$
   35
5. $6 \times 3^2 + 13 - 12$
   $6 \times 9 + 13 - 12$
   $54 + 13 - 12$
   $67 - 12$
   55
6. $32 - 8 \div 1^2 \times 4$
   $32 - 8 \div 1 \times 4$
   $32 - 8 \times 4$
   $32 - 32$
   0
7. $2^2 + 38 - 2 \times 4^2$
   $4 + 38 - 2 \times 4^2$
   $4 + 38 - 2 \times 16$
   $4 + 38 - 32$
   $42 - 32$
   10
8. $7 \times 8 + 6^2 \div 6^2$
   $7 \times 8 + 36 \div 6^2$
   $7 \times 8 + 36 \div 36$
   $56 + 36 \div 36$
   $56 + 1$
   57
9. $19 - 16 \div 1^2 + 2$
   $19 - 16 \div 1 + 2$
   $19 - 16 + 2$
   $3 + 2$
   5
10. $18 \div 3^2 \times 25 - 24$
    $18 \div 9 \times 25 - 24$
    $2 \times 25 - 24$
    $50 - 24$
    26

Day 21

1
$(6+10) \div 8 - 1$
$(16) \div 8 - 1$
$16 \div 8 - 1$
$2 - 1$
1

2
$4 + 4 \times (15 - 5)$
$4 + 4 \times (10)$
$4 + 4 \times 10$
$4 + 40$
44

3
$25 - (6 + 2^2 \times 3)$
$25 - (6 + 4 \times 3)$
$25 - (6 + 12)$
$25 - (18)$
$25 - 18$
7

4
$(4^2 - 9) \times (8 + 2)$
$(16 - 9) \times (8 + 2)$
$(7) \times (8 + 2)$
$7 \times (8 + 2)$
$7 \times (10)$
7×10
70

5
$(22 + 8)2 \div 6$
$(30)2 \div 6$
$30 \times 2 \div 6$
$60 \div 6$
10

6
$20 \div (2 \times 2) + 3$
$20 \div (4) + 3$
$20 \div 4 + 3$
$5 + 3$
8

7
$(4 + 20) \div (3^2 - 3)$
$(24) \div (3^2 - 3)$
$24 \div (3^2 - 3)$
$24 \div (9 - 3)$
$24 \div (6)$
$24 \div 6$
4

8
$14 \div (9 + 7 - 2)$
$14 \div (16 - 2)$
$14 \div (14)$
$14 \div 14$
1

9
$(3 + 5 \times 4)10$
$(3 + 20)10$
$(23)10$
23×10
230

10
$32 \div 8(6 + 4)$
$32 \div 8(10)$
$32 \div 8 \times 10$
$4(10)$
4×10
40

Day 22

1
$6 + 6^2 \div (9 - 3)$
$6 + 6^2 \div (6)$
$6 + 6^2 \div 6$
$6 + 36 \div 6$
$6 + 6$
12

2
$2^2 + 4(30 - 5^2)$
$2^2 + 4(30 - 25)$
$2^2 + 4(5)$
$2^2 + 4 \times 5$
$4 + 4 \times 5$
$4 + 20$
24

3
$100 - (3 \times 5 + 7^2)$
$100 - (3 \times 5 + 49)$
$100 - (15 + 49)$
$100 - (64)$
$100 - 64$
36

4
$(4^2 - 14)^2 (10 - 3^2)$
$(16 - 14)^2 (10 - 3^2)$
$(2)^2 (10 - 3^2)$
$2^2 (10 - 3^2)$
$2^2 (10 - 9)$
$2^2 (1)$
$2^2 \cdot 1$
$4 \cdot 1$
4

5
$(2 + 8)3^2 \div 3$
$(10)3^2 \div 3$
$10 \times 3^2 \div 3$
$10 \times 9 \div 3$
$90 \div 3$
30

6
$5^2 \div (7 - 2) + 14$
$5^2 \div (5) + 14$
$5^2 \div 5 + 14$
$25 \div 5 + 14$
$5 + 14$
19

7
$150 \div (8 - 3)^2 \times 2$
$150 \div (5)^2 \times 2$
$150 \div 5^2 \times 2$
$150 \div 25 \times 2$
6×2
12

8
$18 \div (6^2 - 5 \times 6)$
$18 \div (36 - 5 \times 6)$
$18 \div (36 - 30)$
$18 \div (6)$
$18 \div 6$
3

9
$(8^2 + 3 \times 4)10$
$(64 + 3 \times 4)10$
$(64 + 12)10$
$(76)10$
76×10
760

10
$(8 + 4^2) \div (5 + 3)$
$(8 + 16) \div (5 + 3)$
$(24) \div (5 + 3)$
$24 \div (8)$
$24 \div 8$
3

Day 23

1
$(2 \times 4^2)(27 - 26)$
$(2 \times 16)(27 - 26)$
$(32)(27 - 26)$
$32 \times (1)$
32×1
32

2
$9(3 + 7^2) - 5$
$9(3 + 49) - 5$
$9(52) - 5$
$9 \times 52 - 5$
$468 - 5$
463

3
$2^2 + 6(25 \div 5^2)$
$2^2 + 6(25 \div 25)$
$2^2 + 6(1)$
$2^2 + 6 \times 1$
$4 + 6 \times 1$
$4 + 6$
10

4
$(3^2 + 7 \times 3^2)2$
$(9 + 7 \times 3^2)2$
$(9 + 7 \times 9)2$
$(9 + 63)2$
$(72)2$
$72 \cdot 2$
144

5
$(47 - 46) + 6^2 \times 6$
$(1) + 6^2 \times 6$
$1 + 6^2 \times 6$
$1 + 36 \times 6$
$1 + 216$
217

6
$(10^2 + 10) \div (10 + 1^2)$
$(100 + 10) \div (10 + 1^2)$
$(110) \div (10 + 1^2)$
$110 \div (10 + 1^2)$
$110 \div (10 + 1)$
$110 \div (11)$
$110 \div 11$
10

7
$49 + (45 \div 3^2 - 2^2)$
$49 + (45 \div 9 - 2^2)$
$49 + (45 \div 9 - 4)$
$49 + (5 - 4)$
$49 + (1)$
$49 + 1$
50

8
$(14 \div 7) - 1 \times 1^2$
$(2) - 1 \times 1^2$
$2 - 1 \times 1^2$
$2 - 1 \times 1$
$2 - 1$
1

9
$(18^2 \div 18) - (3^2 + 8)$
$(324 \div 18) - (3^2 + 8)$
$(18) - (3^2 + 8)$
$18 - (3^2 + 8)$
$18 - (9 + 8)$
$18 - (17)$
$18 - 17$
1

10
$35 + (9^2 - 7^2) + 6$
$35 + (81 - 7^2) + 6$
$35 + (81 - 49) + 6$
$35 + (32) + 6$
$35 + 32 + 6$
$67 + 6$
73

Day 24

① $8^2 - (9^2 \div 3) + 5^2$
$8^2 - (81 \div 3) + 5^2$
$8^2 - (27) + 5^2$
$8^2 - 27 + 5^2$
$64 - 27 + 5^2$
$64 - 27 + 25$
$37 + 25$
62

② $(11 + 1^2) \div 3 + 6$
$(11 + 1) \div 3 + 6$
$(12) \div 3 + 6$
$12 \div 3 + 6$
$4 + 6$
10

③ $32 + (4 \times 3^2 \div 2)$
$32 + (4 \times 9 \div 2)$
$32 + (36 \div 2)$
$32 + (18)$
$32 + 18$
50

④ $(42 + 7) - (1^2 \times 7^2)$
$(49) - (1^2 \times 7^2)$
$49 - (1^2 \times 7^2)$
$49 - (1 \times 7^2)$
$49 - (1 \times 49)$
$49 - (49)$
$49 - 49$
0

⑤ $(4^2 - 14 + 4^2) \div 6$
$(16 - 14 + 4^2) \div 6$
$(16 - 14 + 16) \div 6$
$(2 + 16) \div 6$
$(18) \div 6$
$18 \div 6$
3

⑥ $1 \times 5^2 \div (50 - 5^2)$
$1 \times 5^2 \div (50 - 25)$
$1 \times 5^2 \div (25)$
$1 \times 5^2 \div 25$
$1 \times 25 \div 25$
$25 \div 25$
1

⑦ $(34 - 6) \div 2^2 - 2$
$(28) \div 2^2 - 2$
$28 \div 2^2 - 2$
$28 \div 4 - 2$
$7 - 2$
5

⑧ $(10^2 \div 50) + (3 \times 10)$
$(100 \div 50) + (3 \times 10)$
$(2) + (3 \times 10)$
$2 + (3 \times 10)$
$2 + (30)$
$2 + 30$
32

⑨ $36 \div (5^2 - 7 \times 3)$
$36 \div (25 - 7 \times 3)$
$36 \div (25 - 21)$
$36 \div (4)$
$36 \div 4$
9

⑩ $(7 \times 4 - 11 + 12)^2$
$(28 - 11 + 12)^2$
$(17 + 12)^2$
$(29)^2$
29^2
841

Day 25

① $35 \div 7\ (50 - 7^2)$
$35 \div 7\ (50 - 49)$
$35 \div 7\ (1)$
$35 \div 7 \times 1$
5×1
5

② $10(9^2 \div 9^2 - 1)$
$10(81 \div 9^2 - 1)$
$10(81 \div 81 - 1)$
$10(1 - 1)$
$10(0)$
10×0
0

③ $(50 + 3^2 \times 2) \div 2^2$
$(50 + 9 \times 2) \div 2^2$
$(50 + 18) \div 2^2$
$(68) \div 2^2$
$68 \div 2^2$
$68 \div 4$
17

④ $8^2 \div (28 - 26) + 1^2$
$8^2 \div (2) + 1^2$
$8^2 \div 2 + 1^2$
$64 \div 2 + 1^2$
$64 \div 2 + 1$
$32 + 1$
33

⑤ $(18 - 4^2)\ (5 + 12)$
$(18 - 16)\ (5 + 12)$
$(2)\ (5 + 12)$
$2\ (5 + 12)$
$2\ (17)$
2×17
34

⑥ $(50 - 6^2 + 1) - 3$
$(50 - 36 + 1) - 3$
$(14 + 1) - 3$
$(15) - 3$
$15 - 3$
12

⑦ $10(22 \div 11) + 5^2$
$10(2) + 5^2$
$10 \times 2 + 5^2$
$10 \times 2 + 25$
$20 + 25$
45

⑧ $18 - (3^2 \times 9 \div 9)$
$18 - (9 \times 9 \div 9)$
$18 - (81 \div 9)$
$18 - (9)$
$18 - 9$
9

⑨ $(6^2 \div 18 - 2) + 27$
$(36 \div 18 - 2) + 27$
$(2 - 2) + 27$
$(0) + 27$
$0 + 27$
27

⑩ $(50 \times 4) - 11 + 9^2$
$(200) - 11 + 9^2$
$200 - 11 + 9^2$
$200 - 11 + 81$
$189 + 81$
270

Day 26

① $7^2 \times 1 - (22 + 9)$
$7^2 \times 1 - (31)$
$7^2 \times 1 - 31$
$49 \times 1 - 31$
$49 - 31$
18

② $16 \div 8(3 \times 2^2)$
$16 \div 8(3 \times 4)$
$16 \div 8(12)$
$16 \div 8 \times 12$
2×12
24

③ $28 - (4 \times 4 \div 4)^2$
$28 - (16 \div 4)^2$
$28 - (4)^2$
$28 - 4^2$
$28 - 16$
12

④ $(6 + 5)^2 - (1 + 6)^2$
$(11)^2 - (1 + 6)^2$
$11^2 - (1 + 6)^2$
$11^2 - (7)^2$
$11^2 - 7^2$
$121 - 7^2$
$121 - 49$
72

⑤ $(9^2 \times 1^2 \div 3^2) + 10$
$(81 \times 1^2 \div 3^2) + 10$
$(81 \times 1 \div 3^2) + 10$
$(81 \times 1 \div 9) + 10$
$(81 \div 9) + 10$
$(9) + 10$
$9 + 10$
19

⑥ $(38 - 28) \times 6^2 \div 36$
$(10) \times 6^2 \div 36$
$10 \times 6^2 \div 36$
$10 \times 36 \div 36$
$360 \div 36$
10

⑦ $296 - (21 + 10 \times 5^2)$
$296 - (21 + 10 \times 25)$
$296 - (21 + 250)$
$296 - (271)$
$296 - 271$
25

⑧ $12 + (6^2 \div 2^2 - 9)$
$12 + (36 \div 2^2 - 9)$
$12 + (36 \div 4 - 9)$
$12 + (9 - 9)$
$12 + (0)$
$12 + 0$
12

⑨ $7^2 - (8 + 7) \times 3$
$7^2 - (15) \times 3$
$7^2 - 15 \times 3$
$49 - 15 \times 3$
$49 - 45$
4

⑩ $(2 \times 7)\ (18 - 17)$
$(14)\ (18 - 17)$
$14\ (18 - 17)$
$14\ (1)$
14×1
14

Day 27

(1) $(16 - 10)^2 + (13 + 20)$
$(6)^2 + (13 + 20)$
$6^2 + (13 + 20)$
$6^2 + (33)$
$6^2 + 33$
$36 + 33$
69

(2) $2^2 \div (1 + 1 \times 1)$
$2^2 \div (1 + 1)$
$2^2 \div (2)$
$2^2 \div 2$
$4 \div 2$
2

(3) $5 + (18 \div 3)^2 \times 3$
$5 + (6)^2 \times 3$
$5 + 6^2 \times 3$
$5 + 36 \times 3$
$5 + 108$
113

(4) $(16 \times 5^2 - 8) + 70$
$(16 \times 25 - 8) + 70$
$(400 - 8) + 70$
$(392) + 70$
$392 + 70$
462

(5) $(11 + 12)^2 - (12 \div 12)^2$
$(23)^2 - (12 \div 12)^2$
$23^2 - (12 \div 12)^2$
$23^2 - (1)^2$
$23^2 - 1^2$
$529 - 1^2$
$529 - 1$
528

(6) $37 - (20 \times 4) \div 8$
$37 - (80) \div 8$
$37 - 80 \div 8$
$37 - 10$
27

(7) $10^2 \div (5 + 3 \times 5)$
$10^2 \div (5 + 15)$
$10^2 \div (20)$
$10^2 \div 20$
$100 \div 20$
5

(8) $27 + 7(2 \times 1)$
$27 + 7(2)$
$27 + 7 \times 2$
$27 + 14$
41

(9) $30 + (39 - 9) \div 10$
$30 + (30) \div 10$
$30 + 30 \div 10$
$30 + 3$
33

(10) $(15 \times 6 - 40) + 7$
$(90 - 40) + 7$
$(50) + 7$
$50 + 7$
57

Day 28

(1) $29 + 36 - (10^2 \div 50)$
$29 + 36 - (100 \div 50)$
$29 + 36 - (2)$
$29 + 36 - 2$
$65 - 2$
63

(2) $(3^2 - 6) - (1 + 1^2)$
$(9 - 6) - (1 + 1^2)$
$(3) - (1 + 1^2)$
$3 - (1 + 1^2)$
$3 - (1 + 1)$
$3 - 2$
1

(3) $(4 \times 2^2 + 3) - 4^2$
$(4 \times 4 + 3) - 4^2$
$(16 + 3) - 4^2$
$(19) - 4^2$
$19 - 4^2$
$19 - 16$
3

(4) $(81 \div 9 + 6^2) \times 1$
$(81 \div 9 + 36) \times 1$
$(9 + 36) \times 1$
$(45) \times 1$
45×1
45

(5) $(7^2 - 49) + (2 \times 1)^2$
$(49 - 49) + (2 \times 1)^2$
$(0) + (2 \times 1)^2$
$0 + (2 \times 1)^2$
$0 + (2)^2$
$0 + 2^2$
$0 + 4$
4

(6) $(2 + 8)(8 + 9)$
$(10)(8 + 9)$
$10 \times (8 + 9)$
$10 \times (17)$
10×17
170

(7) $10^2 \times (5 - 10 \div 5)$
$10^2 \times (5 - 2)$
$10^2 \times (3)$
$10^2 \times 3$
100×3
300

(8) $(4 + 3 \times 8) \div 7$
$(4 + 24) \div 7$
$(28) \div 7$
$28 \div 7$
4

(9) $6^2 + (6^2 - 4 \div 4)$
$6^2 + (36 - 4 \div 4)$
$6^2 + (36 - 1)$
$6^2 + (35)$
$6^2 + 35$
$36 + 35$
71

(10) $(2^2)(81 \div 3^2) - 20$
$(4)(81 \div 3^2) - 20$
$4(81 \div 3^2) - 20$
$4(81 \div 9) - 20$
$4(9) - 20$
$4 \times 9 - 20$
$36 - 20$
16

Day 29

(1) $3^2 - (1 \times 2^2 + 1^2)$
$3^2 - (1 \times 4 + 1^2)$
$3^2 - (1 \times 4 + 1)$
$3^2 - (4 + 1)$
$3^2 - (5)$
$3^2 - 5$
$9 - 5$
4

(2) $5 \times 9 + (12 - 2^2)$
$5 \times 9 + (12 - 4)$
$5 \times 9 + (8)$
$5 \times 9 + 8$
$45 + 8$
53

(3) $(18 + 8 - 19)^2 - 12$
$(26 - 19)^2 - 12$
$(7)^2 - 12$
$7^2 - 12$
$49 - 12$
37

(4) $(5 - 1)^2 \div (9 - 1)$
$(4)^2 \div (9 - 1)$
$4^2 \div (9 - 1)$
$4^2 \div (8)$
$4^2 \div 8$
$16 \div 8$
2

(5) $11 + (5^2 \div 5^2) - 11$
$11 + (25 \div 5^2) - 11$
$11 + (25 \div 25) - 11$
$11 + (1) - 11$
$11 + 1 - 11$
$12 - 11$
1

(6) $49 \div 7 \times (6^2 - 5^2)$
$49 \div 7 \times (36 - 5^2)$
$49 \div 7 \times (36 - 25)$
$49 \div 7 \times (11)$
$49 \div 7 \times 11$
7×11
77

(7) $5(2 + 2) \times 3^2$
$5(4) \times 3^2$
$5 \times 4 \times 3^2$
$5 \times 4 \times 9$
20×9
180

(8) $16 + (7^2 \times 1^2 - 9)$
$16 + (49 \times 1^2 - 9)$
$16 + (49 \times 1 - 9)$
$16 + (49 - 9)$
$16 + (40)$
$16 + 40$
56

(9) $(9^2 + 3^2 - 50) \div 8$
$(81 + 3^2 - 50) \div 8$
$(81 + 9 - 50) \div 8$
$(90 - 50) \div 8$
$(40) \div 8$
$40 \div 8$
5

(10) $(9 - 2)^2 - (9 \times 2)$
$(7)^2 - (9 \times 2)$
$7^2 - (9 \times 2)$
$7^2 - (18)$
$7^2 - 18$
$49 - 18$
31

Day 30

1. $20 \times 6^2 - (8 + 1)^2$
$20 \times 6^2 - (9)^2$
$20 \times 6^2 - 9^2$
$20 \times 36 - 9^2$
$20 \times 36 - 81$
$720 - 81$
639

2. $160 \div (1 + 3 \times 5)$
$160 \div (1 + 15)$
$160 \div (16)$
$160 \div 16$
10

3. $9 + (3 \times 3 \div 3)^2$
$9 + (9 \div 3)^2$
$9 + (3)^2$
$9 + 3^2$
$9 + 9$
18

4. $(35 + 4^2) - (1 + 2^2)^2$
$(35 + 16) - (1 + 2^2)^2$
$(51) - (1 + 2^2)^2$
$51 - (1 + 2^2)^2$
$51 - (1 + 4)^2$
$51 - (5)^2$
$51 - 5^2$
$51 - 25$
26

5. $(49 - 7^2 + 3) \div 3$
$(49 - 49 + 3) \div 3$
$(0 + 3) \div 3$
$(3) \div 3$
$3 \div 3$
1

6. $(36 \div 6) \times 4 - 2^2$
$(6) \times 4 - 2^2$
$6 \times 4 - 2^2$
$6 \times 4 - 4$
$24 - 4$
20

7. $19 + (8^2 - 6 \times 7)$
$19 + (64 - 6 \times 7)$
$19 + (64 - 42)$
$19 + (22)$
$19 + 22$
41

8. $5^2 + (6 \div 3) - 5^2$
$5^2 + (2) - 5^2$
$5^2 + 2 - 5^2$
$25 + 2 - 5^2$
$25 + 2 - 25$
$27 - 25$
2

9. $(5^2 - 24 + 7) \div 4$
$(25 - 24 + 7) \div 4$
$(1 + 7) \div 4$
$(8) \div 4$
$8 \div 4$
2

10. $(1 \times 6)^2 - (4 - 2)^2$
$(6)^2 - (4 - 2)^2$
$6^2 - (4 - 2)^2$
$6^2 - (2)^2$
$6^2 - 2^2$
$36 - 2^2$
$36 - 4$
32

Day 31

1. $(349 - 7^2)(1 + 2)$
$(349 - 49)(1 + 2)$
$(300)(1 + 2)$
$300(1 + 2)$
$300(3)$
300×3
900

2. $(3 + 2)^2 \div (30 - 5^2)$
$(5)^2 \div (30 - 5^2)$
$5^2 \div (30 - 5^2)$
$5^2 \div (30 - 25)$
$5^2 \div (5)$
$5^2 \div 5$
$25 \div 5$
5

3. $(6^2 - 1^2 + 5^2) \div 30$
$(36 - 1^2 + 5^2) \div 30$
$(36 - 1 + 5^2) \div 30$
$(36 - 1 + 25) \div 30$
$(35 + 25) \div 30$
$(60) \div 30$
$60 \div 30$
2

4. $(9 \div 3^2 + 3^2) \times 4$
$(9 \div 9 + 3^2) \times 4$
$(9 \div 9 + 9) \times 4$
$(1 + 9) \times 4$
$(10) \times 4$
10×4
40

5. $(10 - 8)^2 + (2 - 1)^2$
$(2)^2 + (2 - 1)^2$
$2^2 + (2 - 1)^2$
$2^2 + (1)^2$
$2^2 + 1^2$
$4 + 1^2$
$4 + 1$
5

6. $(6^2 \div 3) - (4 \times 1^2)$
$(36 \div 3) - (4 \times 1^2)$
$(12) - (4 \times 1^2)$
$12 - (4 \times 1^2)$
$12 - (4 \times 1)$
$12 - (4)$
$12 - 4$
8

7. $45 + (6 \times 2^2 + 2^2)$
$45 + (6 \times 4 + 2^2)$
$45 + (6 \times 4 + 4)$
$45 + (24 + 4)$
$45 + (28)$
$45 + 28$
73

8. $50 \div (36 - 26)(7)$
$50 \div (10)(7)$
$50 \div 10(7)$
$50 \div 10 \times 7$
5×7
35

9. $7 + (43 - 14 \div 2)$
$7 + (43 - 7)$
$7 + (36)$
$7 + 36$
43

10. $100(81 \div 9^2 - 1^2)$
$100(81 \div 81 - 1^2)$
$100(81 \div 81 - 1)$
$100(1 - 1)$
$100(0)$
100×0
0

Day 32

1. $(6 + 4)^2 - (5 \div 1)^2$
$(10)^2 - (5 \div 1)^2$
$10^2 - (5 \div 1)^2$
$10^2 - (5)^2$
$10^2 - 5^2$
$100 - 5^2$
$100 - 25$
75

2. $28 - (49 \div 7^2 \times 28)$
$28 - (49 \div 49 \times 28)$
$28 - (1 \times 28)$
$28 - (28)$
$28 - 28$
0

3. $3 \div (2 - 1)^2 \times 3^2$
$3 \div (1)^2 \times 3^2$
$3 \div 1^2 \times 3^2$
$3 \div 1 \times 3^2$
$3 \div 1 \times 9$
3×9
27

4. $(4 \times 2 - 2)^2 + 2$
$(8 - 2)^2 + 2$
$(6)^2 + 2$
$6^2 + 2$
$36 + 2$
38

5. $(5 \times 6 - 2) \div (7 - 5)$
$(30 - 2) \div (7 - 5)$
$(28) \div (7 - 5)$
$28 \div (7 - 5)$
$28 \div (2)$
$28 \div 2$
14

6. $6 + 12 \div (10 - 7)$
$6 + 12 \div (3)$
$6 + 12 \div 3$
$6 + 4$
10

7. $(4)(7^2 + 1^2 \times 3)$
$4(7^2 + 1^2 \times 3)$
$4(49 + 1^2 \times 3)$
$4(49 + 1 \times 3)$
$4(49 + 3)$
$4(52)$
4×52
208

8. $27 - (77 \div 11)(2)$
$27 - (7)(2)$
$27 - 7(2)$
$27 - 7 \times 2$
$27 - 14$
13

9. $80 \div (1 + 3) + 10$
$80 \div (4) + 10$
$80 \div 4 + 10$
$20 + 10$
30

10. $(7 \times 2 - 2) \div 6$
$(14 - 2) \div 6$
$(12) \div 6$
$12 \div 6$
2

Day 33

① $18 \div 9 + (4 - 3)^2$
$18 \div 9 + (1)^2$
$18 \div 9 + 1^2$
$18 \div 9 + 1$
$2 + 1$
3

② $3(1 + 7 \times 2)$
$3(1 + 14)$
$3(15)$
3×15
45

③ $6^2 + (4^2 - 2 \times 3)$
$6^2 + (16 - 2 \times 3)$
$6^2 + (16 - 6)$
$6^2 + (10)$
$6^2 + 10$
$36 + 10$
46

④ $(8 + 0)^2 - (24 + 27)$
$(8)^2 - (24 + 27)$
$8^2 - (24 + 27)$
$8^2 - (51)$
$8^2 - 51$
$64 - 51$
13

⑤ $(35 + 25 - 12) \div 8$
$(60 - 12) \div 8$
$(48) \div 8$
$48 \div 8$
6

⑥ $150 \div 15 \times (4^2 - 2^2)$
$150 \div 15 \times (16 - 2^2)$
$150 \div 15 \times (16 - 4)$
$150 \div 15 \times (12)$
$150 \div 15 \times 12$
10×12
120

⑦ $19 - (3 \div 3 \times 1)^2$
$19 - (1 \times 1)^2$
$19 - (1)^2$
$19 - 1$
18

⑧ $5 + (6 \div 6) - 2^2$
$5 + (1) - 2^2$
$5 + 1 - 2^2$
$5 + 1 - 4$
$6 - 4$
2

⑨ $(3 + 15 \div 3)^2 + 16$
$(3 + 5)^2 + 16$
$(8)^2 + 16$
$8^2 + 16$
$64 + 16$
80

⑩ $(7^2 \times 4^2) + (30 - 21)$
$(49 \times 4^2) + (30 - 21)$
$(49 \times 16) + (30 - 21)$
$(784) + (30 - 21)$
$784 + (30 - 21)$
$784 + (9)$
$784 + 9$
793

Day 34

① $(8 - 8) \times (3^2 + 5^2)$
$(0) \times (3^2 + 5^2)$
$0 \times (3^2 + 5^2)$
$0 \times (9 + 5^2)$
$0 \times (9 + 25)$
$0 \times (34)$
0×34
0

② $28 \div (10 - 3 \times 2)$
$28 \div (10 - 6)$
$28 \div (4)$
$28 \div 4$
7

③ $6 + (9 - 4) \times 7$
$6 + (5) \times 7$
$6 + 5 \times 7$
$6 + 35$
41

④ $(28 - 21 + 1) \times 3^2$
$(7 + 1) \times 3^2$
$(8) \times 3^2$
8×3^2
8×9
72

⑤ $(8^2 + 1^2) - (12 \div 3)$
$(64 + 1^2) - (12 \div 3)$
$(64 + 1) - (12 \div 3)$
$(65) - (12 \div 3)$
$65 - (12 \div 3)$
$65 - (4)$
$65 - 4$
61

⑥ $12 - (6^2 \div 3) \div 2$
$12 - (36 \div 3) \div 2$
$12 - (12) \div 2$
$12 - 12 \div 2$
$12 - 6$
6

⑦ $30 + (5 - 3 \div 1)^2$
$30 + (5 - 3)^2$
$30 + (2)^2$
$30 + 2^2$
$30 + 4$
34

⑧ $2 + 5(2 \times 1)$
$2 + 5(2)$
$2 + 5 \times 2$
$2 + 10$
12

⑨ $6^2 + (5^2 - 7) \div 6$
$6^2 + (25 - 7) \div 6$
$6^2 + (18) \div 6$
$6^2 + 18 \div 6$
$36 + 18 \div 6$
$36 + 3$
39

⑩ $(30 - 10 \times 2) + 7$
$(30 - 20) + 7$
$(10) + 7$
$10 + 7$
17

Day 35

① $12 \div (2 \times 3 - 4)^2$
$12 \div (6 - 4)^2$
$12 \div (2)^2$
$12 \div 2^2$
$12 \div 4$
3

② $30 - 4^2 \div (14 - 6)$
$30 - 4^2 \div (8)$
$30 - 4^2 \div 8$
$30 - 16 \div 8$
$30 - 2$
28

③ $(9^2 + 23 - 49) \div 5$
$(81 + 23 - 49) \div 5$
$(104 - 49) \div 5$
$(55) \div 5$
$55 \div 5$
11

④ $(7 + 2)(1 + 4)$
$(9)(1 + 4)$
$9(1 + 4)$
$9(5)$
9×5
45

⑤ $8^2 + (81 \div 9) - 7^2$
$8^2 + (9) - 7^2$
$8^2 + 9 - 7^2$
$64 + 9 - 7^2$
$64 + 9 - 49$
$73 - 49$
24

⑥ $60 \div 3 \times (28 \div 7)$
$60 \div 3 \times (4)$
$60 \div 3 \times 4$
20×4
80

⑦ $8 \div (6 \div 3)^2 \times 5$
$8 \div (2)^2 \times 5$
$8 \div 2^2 \times 5$
$8 \div 4 \times 5$
2×5
10

⑧ $16 + (3 + 2^2 \times 9)$
$16 + (3 + 4 \times 9)$
$16 + (3 + 36)$
$16 + (39)$
$16 + 39$
55

⑨ $(9^2 + 3^2 - 50) \div 2$
$(81 + 3^2 - 50) \div 2$
$(81 + 9 - 50) \div 2$
$(90 - 50) \div 2$
$(40) \div 2$
$40 \div 2$
20

⑩ $(6^2 - 24) - (3^2 \times 1)$
$(36 - 24) - (3^2 \times 1)$
$(12) - (3^2 \times 1)$
$12 - (3^2 \times 1)$
$12 - (9 \times 1)$
$12 - (9)$
$12 - 9$
3

Day 36

(1)
$((6 + 7^2) \div 5)2$
$((6 + 49) \div 5)2$
$((55) \div 5)2$
$(55 \div 5)2$
$(11)2$
11×2
22

(2)
$7((40 - 6^2) \div 2)$
$7((40 - 36) \div 2)$
$7((4) \div 2)$
$7(4 \div 2)$
$7(2)$
7×2
14

(3)
$4(72 - (4 \times 7 + 6^2))$
$4(72 - (4 \times 7 + 36))$
$4(72 - (28 + 36))$
$4(72 - (64))$
$4(72 - 64)$
$4(8)$
4×8
32

(4)
$(8^2 - 59)((3 + 3^2) \div 2)$
$(64 - 59)((3 + 3^2) \div 2)$
$(5)((3 + 3^2) \div 2)$
$5((3 + 3^2) \div 2)$
$5((3 + 9) \div 2)$
$5((12) \div 2)$
$5(12 \div 2)$
$5(6)$
5×6
30

(5)
$(4 + (6 - 4)) \div 3$
$(4 + (2)) \div 3$
$(4 + 2) \div 3$
$(6) \div 3$
$6 \div 3$
2

(6)
$2^2((7 - 2)2)$
$2^2((5)2)$
$2^2(5 \times 2)$
$2^2(10)$
$2^2 \times 10$
4×10
40

(7)
$36 \div ((3 + 3)^2 \div 4)$
$36 \div ((6)^2 \div 4)$
$36 \div (6^2 \div 4)$
$36 \div (36 \div 4)$
$36 \div (9)$
$36 \div 9$
4

(8)
$56 \div (2^2(5 - 3))$
$56 \div (2^2(2))$
$56 \div (2^2 \times 2)$
$56 \div (4 \times 2)$
$56 \div (8)$
$56 \div 8$
7

(9)
$(61 - (7^2 + 6))10$
$(61 - (49 + 6))10$
$(61 - (55))10$
$(61 - 55)10$
$(6)10$
6×10
60

(10)
$(12 - 4) \div ((6 + 2) \div 4)$
$(8) \div ((6 + 2) \div 4)$
$8 \div ((6 + 2) \div 4)$
$8 \div ((8) \div 4)$
$8 \div (8 \div 4)$
$8 \div (2)$
$8 \div 2$
4

Day 37

(1)
$((31 + 23)) - (9 \div 3)^2$
$((54)) - (9 \div 3)^2$
$(54) - (9 \div 3)^2$
$54 - (9 \div 3)^2$
$54 - (3)^2$
$54 - 3^2$
$54 - 9$
45

(2)
$80 - ((6 + 2)^2 + 2)$
$80 - ((8)^2 + 2)$
$80 - (8^2 + 2)$
$80 - (64 + 2)$
$80 - (66)$
$80 - 66$
14

(3)
$34 - (5^2 + 4^2 \div (1 + 1))$
$34 - (5^2 + 4^2 \div (2))$
$34 - (5^2 + 4^2 \div 2)$
$34 - (25 + 4^2 \div 2)$
$34 - (25 + 16 \div 2)$
$34 - (25 + 8)$
$34 - (33)$
$34 - 33$
1

(4)
$(3^2 - 2) + (7 \div (4 - 3))^2$
$(9 - 2) + (7 \div (4 - 3))^2$
$(7) + (7 \div (4 - 3))^2$
$7 + (7 \div (4 - 3))^2$
$7 + (7 \div (1))^2$
$7 + (7 \div 1)^2$
$7 + (7)^2$
$7 + 7^2$
$7 + 49$
56

(5)
$((4 \times 2 \div 2)^2 - 4) + 4^2$
$((8 \div 2)^2 - 4) + 4^2$
$((4)^2 - 4) + 4^2$
$(4^2 - 4) + 4^2$
$(16 - 4) + 4^2$
$(12) + 4^2$
$12 + 4^2$
$12 + 16$
28

(6)
$2((34 - 25) + 9^2 - 8^2)$
$2((9) + 9^2 - 8^2)$
$2(9 + 9^2 - 8^2)$
$2(9 + 81 - 8^2)$
$2(9 + 81 - 64)$
$2(90 - 64)$
$2(26)$
2×26
52

(7)
$45 - (3 + (6^2 \div 9)^2)$
$45 - (3 + (36 \div 9)^2)$
$45 - (3 + (4)^2)$
$45 - (3 + 4^2)$
$45 - (3 + 16)$
$45 - (19)$
$45 - 19$
26

(8)
$14 + (6^2 + (8 \div 2 + 33))$
$14 + (6^2 + (4 + 33))$
$14 + (6^2 + (37))$
$14 + (6^2 + 37)$
$14 + (36 + 37)$
$14 + (73)$
$14 + 73$
87

(9)
$150 + ((8 - 2)^2 - 3^2)$
$150 + ((6)^2 - 3^2)$
$150 + (6^2 - 3^2)$
$150 + (36 - 3^2)$
$150 + (36 - 9)$
$150 + (27)$
$150 + 27$
177

(10)
$(4 + (24 - 6)) \div 2$
$(4 + (18)) \div 2$
$(4 + 18) \div 2$
$(22) \div 2$
$22 \div 2$
11

Day 38

(1)
$42 - ((8 - 5)^2 + 3)$
$42 - ((3)^2 + 3)$
$42 - (3^2 + 3)$
$42 - (9 + 3)$
$42 - (12)$
$42 - 12$
30

(2)
$120 \div ((20 - 2) + 12) - 4$
$120 \div ((18) + 12) - 4$
$120 \div (18 + 12) - 4$
$120 \div (30) - 4$
$120 \div 30 - 4$
$4 - 4$
0

(3)
$(16 + (5 + 3^2))2$
$(16 + (5 + 9))2$
$(16 + (14))2$
$(16 + 14)2$
$(30)2$
30×2
60

(4)
$36 - ((6 - 3)^2 + (36 - 9))$
$36 - ((3)^2 + (36 - 9))$
$36 - (3^2 + (36 - 9))$
$36 - (3^2 + (27))$
$36 - (3^2 + 27)$
$36 - (9 + 27)$
$36 - (36)$
$36 - 36$
0

(5)
$((7^2 + 10 - 2) + 6) \div 9$
$((49 + 10 - 2) + 6) \div 9$
$((59 - 2) + 6) \div 9$
$((57) + 6) \div 9$
$(57 + 6) \div 9$
$(63) \div 9$
$63 \div 9$
7

(6)
$3((9 + 1) \times 2^2 - 5^2)$
$3((10) \times 2^2 - 5^2)$
$3(10 \times 2^2 - 5^2)$
$3(10 \times 4 - 5^2)$
$3(10 \times 4 - 25)$
$3(40 - 25)$
$3(15)$
3×15
45

(7)
$4^2 \div ((2 \times 2) + 2^2)$
$4^2 \div ((4) + 2^2)$
$4^2 \div (4 + 2^2)$
$4^2 \div (4 + 4)$
$4^2 \div (8)$
$4^2 \div 8$
$16 \div 8$
2

(8)
$5((6^2 + 8 \div 2) - 33)$
$5((36 + 8 \div 2) - 33)$
$5((36 + 4) - 33)$
$5((40) - 33)$
$5(40 - 33)$
$5(7)$
5×7
35

(9)
$42 \div ((7 - 2)^2 - 19)$
$42 \div ((5)^2 - 19)$
$42 \div (5^2 - 19)$
$42 \div (25 - 19)$
$42 \div (6)$
$42 \div 6$
7

(10)
$(5 + (5 - 1)^2) \div (4 + 3)$
$(5 + (4)^2) \div (4 + 3)$
$(5 + 4^2) \div (4 + 3)$
$(5 + 16) \div (4 + 3)$
$(21) \div (4 + 3)$
$21 \div (4 + 3)$
$21 \div (7)$
$21 \div 7$
3

Day 39

(1)
2(42 + 56 ÷ (9 - 2))
2(42 + 56 ÷ (7))
2(42 + 56 ÷ 7)
2(42 + 8)
2(50)
2x 50
100

(2)
$3((7 + 3) \times 5 - 7^2)$
$3((10) \times 5 - 7^2)$
$3(10 \times 5 - 7^2)$
3(10 x 5 - 49)
3(50 - 49)
3(1)
3 x 1
3

(3)
$30 \div ((8 - 5)^2 - 4)$
$30 \div ((3)^2 - 4)$
$30 \div (3^2 - 4)$
30 ÷ (9 - 4)
30 ÷ (5)
30 ÷ 5
6

(4)
$(100 - (5 + 3)^2) \div 9$
$(100 - (8)^2) \div 9$
$(100 - 8^2) \div 9$
(100 - 64) ÷ 9
(36) ÷ 9
36 ÷ 9
4

(5)
10((9 - 3 ÷ 3) - 1)
10((9 - 1) - 1)
10((8) - 1)
10(8 - 1)
10(7)
10 x 7
70

(6)
$9((5 - (1 + 1)^2 - 1))$
$9((5 - (2)^2 - 1))$
$9((5 - 2^2 - 1))$
9((5 - 4 -1))
9((1 -1))
9((0))
9(0)
9 x 0
0

(7)
$12 + (3(2 + 1))^2$
$12 + (3(3))^2$
$12 + (3 \times 3)^2$
$12 + (9)^2$
$12 + 9^2$
12 + 81
93

(8)
$((2 + 4)^2 + 8) \div (7 - 3)$
$((6)^2 + 8) \div (7 - 3)$
$(6^2 + 8) \div (7 - 3)$
(36 + 8) ÷ (7 - 3)
(44) ÷ (7 - 3)
44 ÷ (7 - 3)
44 ÷ (4)
44 ÷ 4
11

(9)
$6 \div (100 \div (7^2 + 1))$
6 ÷ (100 ÷ (49 +1))
6 ÷ (100 ÷ (50))
6 ÷ (100 ÷ 50)
6 ÷ (2)
6 ÷ 2
3

(10)
$3(3^2 + 4(3 + 1))$
$3(3^2 + 4(4))$
$3(3^2 + 4 \times 4)$
3(9 + 4 x 4)
3(9 + 16)
3 x (25)
3 x 25
75

Day 41

(1)
$30 \div (2 + (8 - 4)^2 \div 2)$
$30 \div (2 + (4)^2 \div 2)$
$30 \div (2 + 4^2 \div 2)$
30 ÷ (2 + 16 ÷ 2)
30 ÷ (2 + 8)
30 ÷ (10)
30 ÷ 10
3

(2)
$3^2 + (3^2 - (8 - 5))$
$3^2 + (3^2 - (3))$
$3^2 + (3^2 - 3)$
$3^2 + (9 - 3)$
$3^2 + (6)$
$3^2 + 6$
9 + 6
15

(3)
$(26 + 16) \times 1^2 \div (1 \times 1)^2$
$(42) \times 1^2 \div (1 \times 1)^2$
$42 \times 1^2 \div (1 \times 1)^2$
$42 \times 1^2 \div (1)^2$
$42 \times 1^2 \div 1^2$
$42 \times 1 \div 1^2$
42 x 1 ÷ 1
42 ÷ 1
42

(4)
$((6 \div 2) - 3) + 1^2 - 1^2$
$((3) - 3) + 1^2 - 1^2$
$(3 - 3) + 1^2 - 1^2$
$(0) + 1^2 - 1^2$
$0 + 1^2 - 1^2$
$0 + 1 - 1^2$
0 +1 - 1
1 - 1
0

(5)
$(8 - (3 - 2))^2 + 37$
$(8 - (1))^2 + 37$
$(8 - 1)^2 + 37$
$(7)^2 + 37$
$7^2 + 37$
49 + 37
86

(6)
120 ÷ ((20 - 2) + 12) -1
120 ÷ ((18) + 12) -1
120 ÷ (18 + 12) -1
120 ÷ (30) -1
120 ÷ 30 -1
4 -1
3

Day 40

(1)
$(32 - 5^2)((20 + 28) \div 8)$
(32 - 25) ((20 + 28) ÷ 8)
(7) ((20 + 28) ÷ 8)
7 ((20 + 28) ÷ 8)
7 ((48) ÷ 8)
7 (48 ÷ 8)
7 (6)
7 x 6
42

(2)
(1 + 7(13 - 8)) ÷ 3
(1 + 7(5)) ÷ 3
(1 + 7 x 5) ÷ 3
(1 + 35) ÷ 3
(36) ÷ 3
36 ÷ 3
12

(3)
$5^2 + (4 \div (3 - 2))^2$
$5^2 + (4 \div (1))^2$
$5^2 + (4 \div 1)^2$
$5^2 + (4)^2$
$5^2 + 4^2$
$25 + 4^2$
25 + 16
41

(4)
((12 - 4 ÷ 2) - 2)3
((12 - 2) - 2)3
((10) - 2)3
(10 - 2)3
(8)3
8 x 3
24

(5)
$((8 - 5)2)^2 - 32$
$((3)2)^2 - 32$
$(3 \times 2)^2 - 32$
$(6)^2 - 32$
$6^2 - 32$
36 - 32
4

(6)
$28 \div ((11 - 9) + (2^2 + 1))$
$28 \div ((2) + (2^2 + 1))$
$28 \div (2 + (2^2 + 1))$
28 ÷ (2 + (4 + 1))
28 ÷ (2 + (5))
28 ÷ (2 + 5)
28 ÷ (7)
28 ÷ 7
4

(7)
$(7^2 - 4) \div ((8 + 7) \div 3)$
(49 - 4) ÷ ((8 + 7) ÷ 3)
(45) ÷ ((8 + 7) ÷ 3)
45 ÷ ((8 + 7) ÷ 3)
45 ÷ ((15) ÷ 3)
45 ÷ (15 ÷ 3)
45 ÷ (5)
45 ÷ 5
9

(8)
$6(9^2 - (8 \times 7 + 4^2))$
$6(9^2 - (8 \times 7 + 16))$
$6(9^2 - (56 + 16))$
$6(9^2 - (72))$
$6(9^2 - 72)$
6(81 - 72)
6(9)
6 x 9
54

(9)
$9 + (8^2 - 2(10 + 15))$
$9 + (8^2 - 2(25))$
$9 + (8^2 - 2 \times 25)$
9 + (64 - 2 x 25)
9 + (64 - 50)
9 + (14)
9 + 14
23

(10)
(50 - (5 + 6 x 2))10
(50 - (5 + 12))10
(50 - (17))10
(50 - 17)10
(33)10
33 x 10
330

---

(7)
(5 + 10 - (2+6))9
(5 + 10 - (8))9
(5 + 10 - 8)9
(15 - 8)9
(7)9
7 x 9
63

(8)
((6+6) ÷ 2) (2 + 1)
((12) ÷ 2) (2 + 1)
(12 ÷ 2) (2 + 1)
(6) (2 + 1)
6 (2 + 1)
6 (3)
6 x 3
18

(9)
$(4^2 + (38 - 18)) + 4^2$
$(4^2 + (20)) + 4^2$
$(4^2 + 20) + 4^2$
$(16 + 20) + 4^2$
$(36) + 4^2$
$36 + 4^2$
36 + 16
52

(10)
(50 - (45 - 40)) ÷ (7 - 2)
(50 - (5)) ÷ (7 - 2)
(50 - 5) ÷ (7 - 2)
(45) ÷ (7 - 2)
45 ÷ (7 - 2)
45 ÷ (5)
45 ÷ 5
9

Day 42

1
1 - (3^2 ÷ (14 + 20 - 25))
1 - (3^2 ÷ (34 - 25))
1 - (3^2 ÷ (9))
1 - (3^2 ÷ 9)
1 - (9 ÷ 9)
1 - (1)
1 - 1
0

2
5((7^2 + 5 x 3) - 33)
5((49 + 5 x 3) - 33)
5((49 + 15) - 33)
5((64) - 33)
5(64 - 33)
5(31)
5 x 31
155

3
(55 - 10) + (1^2 x (23 - 13))
(45) + (1^2 x (23 - 13))
45 + (1^2 x (23 - 13))
45 + (1^2 x (10))
45 + (1^2 x 10)
45 + (1 x 10)
45 + (10)
45 + 10
55

4
(8 - (1 + 3))^2 ÷ 8
(8 - (4))^2 ÷ 8
(8 - 4)^2 ÷ 8
(4)^2 ÷ 8
4^2 ÷ 8
16 ÷ 8
2

5
(6 - 1) + ((1 + 2) + 26)
(5) + ((1 + 2) + 26)
5 + ((1 + 2) + 26)
5 + ((3) + 26)
5 + (3 + 26)
5 + (29)
5 + 29
34

6
4^2 ÷ ((12 - 5 x 2) + 6) x 15
4^2 ÷ ((12 - 10) + 6) x 15
4^2 ÷ ((2) + 6) x 15
4^2 ÷ (2 + 6) x 15
4^2 ÷ (8) x 15
4^2 ÷ 8 x 15
16 ÷ 8 x 15
2 x 15
30

7
(80 + (4^2 ÷ 8)) - 6
(80 + (16 ÷ 8)) - 6
(80 + (2)) - 6
(80 + 2) - 6
(82) - 6
82 - 6
76

8
2(42 + 56 ÷ (9 - 2))
2(42 + 56 ÷ (7))
2(42 + 56 ÷ 7)
2(42 + 8)
2(50)
2 x 50
100

9
(6 + (6 - 5))^2 + 8
(6 + (1))^2 + 8
(6 + 1)^2 + 8
(7)^2 + 8
7^2 + 8
49 + 8
57

10
2((70 - 55) + 5^2 - 10)
2((15) + 5^2 - 10)
2(15 + 5^2 - 10)
2(15 + 25 - 10)
2(40 - 10)
2(30)
2 x 30
60

Day 43

1
(5 + (2 x 14) + (5 x 3^2))
(5 + (28) + (5 x 3^2))
(5 + 28 + (5 x 3^2))
(5 + 28 + (5 x 9))
(5 + 28 + (45))
(5 + 28 + 45)
(33 + 45)
(78)
78

2
6^2 ÷ ((7 - 3) 3 - 3)
6^2 ÷ ((4) 3 - 3)
6^2 ÷ (4 x 3 - 3)
6^2 ÷ (12 - 3)
6^2 ÷ (9)
6^2 ÷ 9
36 ÷ 9
4

3
(5 x (48 ÷ 4))2 - 15
(5 x (12))2 - 15
(5 x 12)2 - 15
(60)2 - 15
60 x 2 - 15
120 - 15
105

4
(56 ÷ 7)^2 - ((6 ÷ 6) + 5^2)
(8)^2 - ((6 ÷ 6) + 5^2)
8^2 - ((6 ÷ 6) + 5^2)
8^2 - ((1) + 5^2)
8^2 - (1 + 5^2)
8^2 - (1 + 25)
8^2 - (26)
8^2 - 26
64 - 26
38

5
100 - 50 + (150 - (50+25))
100 - 50 + (150 - (75))
100 - 50 + (150 - 75)
100 - 50 + (75)
100 - 50 + 75
50 + 75
125

6
(22 + (8 x 2^2)) - (4^2 + 16)
(22 + (8 x 4)) - (4^2 + 16)
(22 + (32)) - (4^2 + 16)
(22 + 32) - (4^2 + 16)
(54) - (4^2 + 16)
54 - (4^2 + 16)
54 - (16 + 16)
54 - (32)
54 - 32
22

7
(100 - (4^2 + 9 x 8))2
(100 - (16 + 9 x 8))2
(100 - (16 + 72))2
(100 - (88))2
(100 - 88)2
(12)2
12 x 2
24

8
(3 x 2)^2 - (45 ÷ (7 - 2))
(6)^2 - (45 ÷ (7 - 2))
6^2 - (45 ÷ (7 - 2))
6^2 - (45 ÷ (5))
6^2 - (45 ÷ 5)
6^2 - (9)
6^2 – 9
36- 9
27

9
(4 + (30 - 12)) ÷ 2
(4 + (18)) ÷ 2
(4 + 18) ÷ 2
(22) ÷ 2
22 ÷ 2
11

10
2((20 - 14) + 7 - 3^2)
2((6) + 7 - 3^2)
2(6 + 7 - 3^2)
2(6 + 7 - 9)
2(13 - 9)
2(4)
2 x 4
8

Day 44

1
56 ÷ ((6 - 2)^2 ÷ 2)
56 ÷ ((4)^2 ÷ 2)
56 ÷ (4^2 ÷ 2)
56 ÷ (16 ÷ 2)
56 ÷ (8)
56 ÷ 8
7

2
((3 x 2)^2 - 16) ÷ 5
((6)^2 - 16) ÷ 5
(6^2 - 16) ÷ 5
(36 - 16) ÷ 5
(20) ÷ 5
20 ÷ 5
4

3
23 + ((27 ÷ 9) x 5^2) - 8^2
23 + ((3) x 5^2) - 8^2
23 + (3 x 5^2) - 8^2
23 + (3 x 25) - 8^2
23 + (75) - 8^2
23 + 75 - 8^2
23 + 75 - 64
98 - 64
34

4
(150 ÷ 50) x (4 +(27 - 23))
(3) x (4 +(27 - 23))
3 x (4 +(27 - 23))
3 x (4 +(4))
3 x (4 + 4)
3 x (8)
3 x 8
24

5
((3^2 - 6) x (4 + 2^2)) - 24
((9 - 6) x (4 + 2^2)) - 24
((3) x (4 + 2^2)) – 24
(3 x (4 + 2^2)) – 24
(3 x (4 + 4)) – 24
(3 x (8)) – 24
(3 x 8) – 24
(24) – 24
24 - 24
0

6
6^2 ÷ (16 ÷ (3 + 1))
6^2 ÷ (16 ÷ (4))
6^2 ÷ (16 ÷ 4)
6^2 ÷ (4)
6^2 ÷ 4
36 ÷ 4
9

7
(61 + (8 - 6))10
(61 + (2))10
(61 + 2)10
(63)10
63 x 10
630

8
9 x 3 + (80 ÷ (16 - 14))
9 x 3 + (80 ÷ (2))
9 x 3 + (80 ÷ 2)
9 x 3 + (40)
9 x 3 + 40
27 + 40
67

9
10^2 + ((32 + 68) ÷ 4)
10^2 + ((100) ÷ 4)
10^2 + (100 ÷ 4)
10^2 + (25)
10^2 + 25
100 + 25
125

10
(72 - (4^2 + 4 x 7)) ÷ 2
(72 - (16 + 4 x 7)) ÷ 2
(72 - (16 + 28)) ÷ 2
(72 - (44)) ÷ 2
(72 - 44) ÷ 2
(28) ÷ 2
28 ÷ 2
14

Day 45

1
$44 - 28 + ((8 \div 2)^2 - 13)$
$44 - 28 + ((4)^2 - 13)$
$44 - 28 + (4^2 - 13)$
$44 - 28 + (16 - 13)$
$44 - 28 + (3)$
$44 - 28 + 3$
$16 + 3$
19

2
$50 - ((12 - 7)^2 + 6)$
$50 - ((5)^2 + 6)$
$50 - (5^2 + 6)$
$50 - (25 + 6)$
$50 - (31)$
$50 - 31$
19

3
$(10^2 \div (37 - 17)) + 5 \times 8$
$(10^2 \div (20)) + 5 \times 8$
$(10^2 \div 20) + 5 \times 8$
$(100 \div 20) + 5 \times 8$
$(5) + 5 \times 8$
$5 + 5 \times 8$
$5 + 40$
45

4
$90 + ((3 \div 1 + 6)^2 + 2)$
$90 + ((3 + 6)^2 + 2)$
$90 + ((9)^2 + 2)$
$90 + (9^2 + 2)$
$90 + (81 + 2)$
$90 + (83)$
$90 + 83$
173

5
$9 \times 6 + (8 - (3 + 4))$
$9 \times 6 + (8 - (7))$
$9 \times 6 + (8 - 7)$
$9 \times 6 + (1)$
$9 \times 6 + 1$
$54 + 1$
55

6
$48 \div ((18 - 4^2) + (25 - 15))$
$48 \div ((18 - 16) + (25 - 15))$
$48 \div ((2) + (25 - 15))$
$48 \div (2 + (25 - 15))$
$48 \div (2 + (10))$
$48 \div (2 + 10)$
$48 \div (12)$
$48 \div 12$
4

7
$((20 + 10 - 2) + 8) \div 6$
$((30 - 2) + 8) \div 6$
$((28) + 8) \div 6$
$(28 + 8) \div 6$
$(36) \div 6$
$36 \div 6$
6

8
$(6 + (12 + 7))5$
$(6 + (19))5$
$(6 + 19)5$
$(25)5$
25×5
125

9
$(7 - 4)\ ((8 + 16) \div 2)$
$(3)\ ((8 + 16) \div 2)$
$3\ ((8 + 16) \div 2)$
$3\ ((24) \div 2)$
$3\ (24 \div 2)$
$3\ (12)$
3×12
36

10
$10(2 + (17 - 9))$
$10(2 + (8))$
$10(2 + 8)$
$10(10)$
10×10
100

Day 46

1
$2 + (6 + (3 \times 2)^2) \div 7$
$2 + (6 + (6)^2) \div 7$
$2 + (6 + 6^2) \div 7$
$2 + (6 + 36) \div 7$
$2 + (42) \div 7$
$2 + 42 \div 7$
$2 + 6$
8

2
$(12 - (1 + 7))^2 \div 2$
$(12 - (8))^2 \div 2$
$(12 - 8)^2 \div 2$
$(4)^2 \div 2$
$4^2 \div 2$
$16 \div 2$
8

3
$100 - ((45 + 40) - (4 \times 2)^2)$
$100 - ((85) - (4 \times 2)^2)$
$100 - (85 - (4 \times 2)^2)$
$100 - (85 - (8)^2)$
$100 - (85 - 8^2)$
$100 - (85 - 64)$
$100 - 21$
79

4
$(14 - 6) - (1 + (27 - 22))$
$(8) - (1 + (27 - 22))$
$8 - (1 + (27 - 22))$
$8 - (1 + (5))$
$8 - (1 + 5)$
$8 - (6)$
$8 - 6$
2

5
$(7^2 + (24 - 13)) + 3^2$
$(7^2 + (11)) + 3^2$
$(7^2 + 11) + 3^2$
$(49 + 11) + 3^2$
$(60) + 3^2$
$60 + 9$
69

6
$((15 + 20) \div 5)\ (3 + 4)$
$((35) \div 5)\ (3 + 4)$
$(35 \div 5)\ (3 + 4)$
$(7)\ (3 + 4)$
$7\ (3 + 4)$
7×7
49

7
$(3 + 2)^2 - (10 - (9 - 3))^2$
$(5)^2 - (10 - (9 - 3))^2$
$5^2 - (10 - (9 - 3))^2$
$5^2 - (10 - (6))^2$
$5^2 - (10 - 6)^2$
$5^2 - (4)^2$
$5^2 - 4^2$
$25 - 16$
9

8
$(8 - (3 - 2))^2 + 37$
$(8 - (1))^2 + 37$
$(8 - 1)^2 + 37$
$(7)^2 + 37$
$7^2 + 37$
$49 + 37$
86

9
$8^2 - 1^2 + (24 \div (7 - 1))$
$8^2 - 1^2 + (24 \div (6))$
$8^2 - 1^2 + (24 \div 6)$
$8^2 - 1^2 + (4)$
$8^2 - 1^2 + 4$
$64 - 1^2 + 4$
$64 - 1 + 4$
$63 + 4$
67

10
$90 \div ((38 + 41) - 7^2)$
$90 \div ((79) - 7^2)$
$90 \div (79 - 7^2)$
$90 \div (79 - 49)$
$90 \div (30)$
$90 \div 30$
3

Day 47

1
$(40 - (30 - 8)) \div (7 - 4)$
$(40 - (22)) \div (7 - 4)$
$(40 - 22) \div (7 - 4)$
$(18) \div (7 - 4)$
$18 \div (7 - 4)$
$18 \div (3)$
$18 \div 3$
6

2
$48 \div (1 \times (4^2 - 10))$
$48 \div (1 \times (16 - 10))$
$48 \div (1 \times (6))$
$48 \div (1 \times 6)$
$48 \div (6)$
$48 \div 6$
8

3
$(5^2 + (3 + 3))^2 - (2 \times 2)^2$
$(5^2 + (6))^2 - (2 \times 2)^2$
$(5^2 + 6)^2 - (2 \times 2)^2$
$(25 + 6)^2 - (2 \times 2)^2$
$(31)^2 - (2 \times 2)^2$
$31^2 - (2 \times 2)^2$
$31^2 - (4)^2$
$31^2 - 4^2$
$961 - 4^2$
$961 - 16$
945

4
$4 + (43 - (2 \times 3)^2 \div 6^2)$
$4 + (43 - (6)^2 \div 6^2)$
$4 + (43 - 6^2 \div 6^2)$
$4 + (43 - 36 \div 6^2)$
$4 + (43 - 36 \div 36)$
$4 + (43 - 1)$
$4 + (42)$
$4 + 42$
46

5
$((22 - 17) \times 3) - (18 \div 9)$
$((5) \times 3) - (18 \div 9)$
$(5 \times 3) - (18 \div 9)$
$(15) - (18 \div 9)$
$15 - (18 \div 9)$
$15 - (2)$
$15 - 2$
13

6
$(1 + 2)\ (5 \times (5 - 4))$
$(3)\ (5 \times (5 - 4))$
$3\ (5 \times (5 - 4))$
$3\ (5 \times (1))$
$3\ (5 \times 1)$
$3\ (5)$
3×5
15

7
$4^2 + (5^2 - (20 - 5))$
$4^2 + (5^2 - (15))$
$4^2 + (5^2 - 15)$
$4^2 + (25 - 15)$
$4^2 + (10)$
$4^2 + 10$
$16 + 10$
26

8
$50 - (10 + 2(10 \div 10))$
$50 - (10 + 2(1))$
$50 - (10 + 2 \times 1)$
$50 - (10 + 2)$
$50 - (12)$
$50 - 12$
38

9
$20 + ((3^2 - 2)^2 - (4 \times 2))$
$20 + ((9 - 2)^2 - (4 \times 2))$
$20 + ((7)^2 - (4 \times 2))$
$20 + (7^2 - (4 \times 2))$
$20 + (7^2 - (8))$
$20 + (7^2 - 8)$
$20 + (49 - 8)$
$20 + (41)$
$20 + 41$
61

10
$(40 - 2(4 + 3^2)) + 34$
$(40 - 2(4 + 9)) + 34$
$(40 - 2(13)) + 34$
$(40 - 2 \times 13) + 34$
$(40 - 26) + 34$
$(14) + 34$
$14 + 34$
48

Day 48

(1)
$100 - ((3 \times 4 \div 2)^2 + 4^2)$
$100 - ((12 \div 2)^2 + 4^2)$
$100 - ((6)^2 + 4^2)$
$100 - (6^2 + 4^2)$
$100 - (36 + 4^2)$
$100 - (36 + 16)$
$100 - (52)$
$100 - 52$
48

(2)
$4((5 + 3) \div 2) - 9$
$4((8) \div 2) - 9$
$4(8 \div 2) - 9$
$4(4) - 9$
$4 \times 4 - 9$
$16 - 9$
7

(3)
$(81 \div (4^2 - (2^2 \times 3) + 5))$
$(81 \div (4^2 - (4 \times 3) + 5))$
$(81 \div (4^2 - (12) + 5))$
$(81 \div (4^2 - 12 + 5))$
$(81 \div (16 - 12 + 5))$
$(81 \div (4 + 5))$
$(81 \div (9))$
$(81 \div 9)$
(9)
9

(4)
$(20 - 12) ((3 + 7) \div 2)$
$(8) ((3 + 7) \div 2)$
$8 ((3 + 7) \div 2)$
$8 ((10) \div 2)$
$8 (10 \div 2)$
$8 (5)$
8×5
40

(5)
$22 + ((6 \div 2)^2 - (1 + 5))$
$22 + ((3)^2 - (1 + 5))$
$22 + (3^2 - (1 + 5))$
$22 + (3^2 - (6))$
$22 + (3^2 - 6)$
$22 + (9 - 6)$
$22 + (3)$
$22 + 3$
25

(6)
$(52 - (13 - 6)) \div (8 + 1^2)$
$(52 - (7)) \div (8 + 1^2)$
$(52 - 7) \div (8 + 1^2)$
$(45) \div (8 + 1^2)$
$45 \div (8 + 1^2)$
$45 \div (8 + 1)$
$45 \div (9)$
$45 \div 9$
5

(7)
$3 + (5^2 - (5 + 40 \div 4))$
$3 + (5^2 - (5 + 10))$
$3 + (5^2 - (15))$
$3 + (5^2 - 15)$
$3 + (25 - 15)$
$3 + (10)$
$3 + 10$
13

(8)
$(8 - (3 - 2))^2 + 37$
$(8 - (1))^2 + 37$
$(8 - 1)^2 + 37$
$(7)^2 + 37$
$7^2 + 37$
$49 + 37$
86

(9)
$24 \div ((6 - 4 + 2) \times 2)$
$24 \div ((2 + 2) \times 2)$
$24 \div ((4) \times 2)$
$24 \div (4 \times 2)$
$24 \div (8)$
$24 \div 8$
3

(10)
$((1 + 3) \times 2)^2 - 34$
$((4) \times 2)^2 - 34$
$(4 \times 2)^2 - 34$
$(8)^2 - 34$
$8^2 - 34$
$64 - 34$
30

Day 49

(1)
$((20 + 70) - (3 + 9 \times 8)) \div 3$
$((90) - (3 + 9 \times 8)) \div 3$
$(90 - (3 + 9 \times 8)) \div 3$
$(90 - (3 + 72)) \div 3$
$(90 - (75)) \div 3$
$(90 - 75) \div 3$
$(15) \div 3$
$15 \div 3$
5

(2)
$(5 + 10 - (2 + 6))9$
$(5 + 10 - (8))9$
$(5 + 10 - 8)9$
$(15 - 8)9$
$(7)9$
7×9
63

(3)
$(41 - (14 - 5)) \div (12 - 8)$
$(41 - (9)) \div (12 - 8)$
$(41 - 9) \div (12 - 8)$
$(32) \div (12 - 8)$
$32 \div (12 - 8)$
$32 \div (4)$
$32 \div 4$
8

(4)
$(9^2 - (5 + 9 \times 7)) + 4^2$
$(9^2 - (5 + 63)) + 4^2$
$(9^2 - (68)) + 4^2$
$(9^2 - 68) + 4^2$
$(81 - 68) + 4^2$
$(13) + 4^2$
$13 + 4^2$
$13 + 16$
29

(5)
$(2 \times (6 - 1))^2 \div 4$
$(2 \times (5))^2 \div 4$
$(2 \times 5)^2 \div 4$
$(10)^2 \div 4$
$10^2 \div 4$
$100 \div 4$
25

(6)
$(18 + 22) \div (4 \times (34 - 24))$
$(40) \div (4 \times (34 - 24))$
$40 \div (4 \times (34 - 24))$
$40 \div (4 \times (10))$
$40 \div (4 \times 10)$
$40 \div (40)$
$40 \div 40$
1

(7)
$4^2 \div ((5 \times 7 + 1) - 20)$
$4^2 \div ((35 + 1) - 20)$
$4^2 \div ((36) - 20)$
$4^2 \div (36 - 20)$
$4^2 \div (16)$
$4^2 \div 16$
$16 \div 16$
1

(8)
$((10 - 5) \times ((9 \div 3)^2 + 1))$
$((5) \times ((9 \div 3)^2 + 1))$
$(5 \times ((9 \div 3)^2 + 1))$
$(5 \times ((3)^2 + 1))$
$(5 \times (3^2 + 1))$
$(5 \times (9 + 1))$
$(5 \times (10))$
(5×10)
(50)
50

(9)
$2((2 + 2 - 2)^2 \div 2)$
$2((4 - 2)^2 \div 2)$
$2((2)^2 \div 2)$
$2(2^2 \div 2)$
$2(4 \div 2)$
$2(2)$
2×2
4

(10)
$((50 - 5 \times 8) + 30)2$
$((50 - 40) + 30)2$
$((10) + 30)2$
$(10 + 30)2$
$(40)2$
40×2
80

Day 50

(1)
$(4 \times 5 - (2 + 14)) (5 + 2)$
$(4 \times 5 - (16)) (5 + 2)$
$(4 \times 5 - 16) (5 + 2)$
$(20 - 16) (5 + 2)$
$(4) (5 + 2)$
$4 (5 + 2)$
$4 (7)$
4×7
28

(2)
$7((4^2 + 10 \div 2) - 10)$
$7((16 + 10 \div 2) - 10)$
$7((16 + 5) - 10)$
$7((21) - 10)$
$7(21 - 10)$
$7(11)$
7×11
77

(3)
$120 \div ((6 + 3 \times 3)2)$
$120 \div ((6 + 9)2)$
$120 \div ((15)2)$
$120 \div (15 \times 2)$
$120 \div (30)$
$120 \div 30$
4

(4)
$37 - (6 \div 3 \times (3 + 1^2)^2)$
$37 - (6 \div 3 \times (3 + 1)^2)$
$37 - (6 \div 3 \times (4)^2)$
$37 - (6 \div 3 \times 4^2)$
$37 - (6 \div 3 \times 16)$
$37 - (2 \times 16)$
$37 - (32)$
$37 - 32$
5

(5)
$((7 - 5)^2 + (5 \times 7)) \div 3$
$((2)^2 + (5 \times 7)) \div 3$
$(2^2 + (5 \times 7)) \div 3$
$(2^2 + (35)) \div 3$
$(2^2 + 35) \div 3$
$(4 + 35) \div 3$
$(39) \div 3$
$39 \div 3$
13

(6)
$(2(42 - 24)) + 8 \div 4$
$(2(18)) + 8 \div 4$
$(2 \times 18) + 8 \div 4$
$(36) + 8 \div 4$
$36 + 8 \div 4$
$36 + 2$
38

(7)
$81 \times 2 - ((6 \times 2)^2 + 6)$
$81 \times 2 - ((12)^2 + 6)$
$81 \times 2 - (12^2 + 6)$
$81 \times 2 - (144 + 6)$
$81 \times 2 - (150)$
$81 \times 2 - 150$
$162 - 150$
12

(8)
$43 - ((30 - 9) \div 3 + 2)$
$43 - ((21) \div 3 + 2)$
$43 - (21 \div 3 + 2)$
$43 - (7 + 2)$
$43 - (9)$
$43 - 9$
34

(9)
$(9 + 1)^2 - (15 - (7 - 4))$
$(10)^2 - (15 - (7 - 4))$
$10^2 - (15 - (7 - 4))$
$10^2 - (15 - (3))$
$10^2 - (15 - 3)$
$10^2 - (12)$
$10^2 - 12$
$100 - 12$
88

(10)
$27 + (12 + 30 \div (3 - 2))$
$27 + (12 + 30 \div (1))$
$27 + (12 + 30 \div 1)$
$27 + (12 + 30)$
$27 + (42)$
$27 + 42$
69

Day 51

① $(100 + 50) \div 2 - 3(5-1)$
$(150) \div 2 - 3(5-1)$
$150 \div 2 - 3(5-1)$
$150 \div 2 - 3(4)$
$150 \div 2 - 3 \times 4$
$75 - 3 \times 4$
$75 - 12$
63

② $(3 + 5) \times 2(40 - 4) \div 3^2$
$(8) \times 2(40 - 4) \div 3^2$
$8 \times 2(40 - 4) \div 3^2$
$8 \times 2(36) \div 3^2$
$8 \times 2 \times 36 \div 3^2$
$8 \times 2 \times 36 \div 9$
$16 \times 36 \div 9$
$576 \div 9$
64

③ $30 - 6 + 4^2 - 3 \times 7 + 10$
$30 - 6 + 16 - 3 \times 7 + 10$
$30 - 6 + 16 - 21 + 10$
$24 + 16 - 21 + 10$
$40 - 21 + 10$
$19 + 10$
29

④ $(6^2 - 9 \times 3)(8 + 2 \div 2)$
$(36 - 9 \times 3)(8 + 2 \div 2)$
$(36 - 27)(8 + 2 \div 2)$
$(9)(8 + 2 \div 2)$
$9(8 + 2 \div 2)$
$9(8 + 1)$
$9(9)$
9×9
81

⑤ $(22 + 8)5 \div 25 - 2 \times 2$
$(30)5 \div 25 - 2 \times 2$
$30 \times 5 \div 25 - 2 \times 2$
$150 \div 25 - 2 \times 2$
$6 - 2 \times 2$
$6 - 4$
2

⑥ $24 \div (40 - 4^2 \times 2)(3 + 7)$
$24 \div (40 - 16 \times 2)(3 + 7)$
$24 \div (40 - 32)(3 + 7)$
$24 \div (8)(3 + 7)$
$24 \div 8(3 + 7)$
$24 \div 8(10)$
$24 \div 8 \times 10$
3×10
30

⑦ $(37 + 5) \div (3^2 - 4 \times 4 \div 8)$
$(42) \div (3^2 - 4 \times 4 \div 8)$
$42 \div (3^2 - 4 \times 4 \div 8)$
$42 \div (9 - 4 \times 4 \div 8)$
$42 \div (9 - 16 \div 8)$
$42 \div (9 - 2)$
$42 \div (7)$
$42 \div 7$
6

⑧ $14 \div 2 + 3 \times 8 - 20 \div 4$
$7 + 3 \times 8 - 20 \div 4$
$7 + 24 - 20 \div 4$
$7 + 24 - 5$
$7 + 19$
26

Day 52

① $5 - (1 \times 2) + 3 \div (2 - 1)$
$5 - (2) + 3 \div (2 - 1)$
$5 - 2 + 3 \div (2 - 1)$
$5 - 2 + 3 \div (1)$
$5 - 2 + 3 \div 1$
$5 - 2 + 3$
$3 + 3$
6

② $4 \div 1 \times 3^2 \div (2 + 7) - 3$
$4 \div 1 \times 3^2 \div (9) - 3$
$4 \div 1 \times 3^2 \div 9 - 3$
$4 \div 1 \times 9 \div 9 - 3$
$4 \times 9 \div 9 - 3$
$36 \div 9 - 3$
$4 - 3$
1

③ $40 + 27 - (5 \times 3) + (8 \times 1^2)$
$40 + 27 - (15) + (8 \times 1^2)$
$40 + 27 - 15 + (8 \times 1^2)$
$40 + 27 - 15 + (8 \times 1)$
$40 + 27 - 15 + (8)$
$40 + 27 - 15 + 8$
$67 - 15 + 8$
$52 + 8$
60

④ $(4 \times 2 + 2)^2 \div 20 - 1 + 1$
$(8 + 2)^2 \div 20 - 1 + 1$
$(10)^2 \div 20 - 1 + 1$
$10^2 \div 20 - 1 + 1$
$100 \div 20 - 1 + 1$
$5 - 1 + 1$
$4 + 1$
5

⑤ $9 - 9 + 1 \times 2(4 \times 3)^2$
$9 - 9 + 1 \times 2(12)^2$
$9 - 9 + 1 \times 2 \times 12^2$
$9 - 9 + 1 \times 2 \times 144$
$9 - 9 + 1 \times 288$
$9 - 9 + 288$
$0 + 288$
288

⑥ $(8 + 7 \times 4 - 2) - 3^2 \div 9$
$(8 + 28 - 2) - 3^2 \div 9$
$(36 - 2) - 3^2 \div 9$
$(34) - 3^2 \div 9$
$34 - 3^2 \div 9$
$34 - 9 \div 9$
$34 - 1$
33

⑦ $1 \times 6 - 2 + (80 \div 40 + 3)$
$1 \times 6 - 2 + (2 + 3)$
$1 \times 6 - 2 + (5)$
$1 \times 6 - 2 + 5$
$6 - 2 + 5$
$4 + 5$
9

⑧ $7 \div (6 - 5) + 28 \times (5 \div 5)^2$
$7 \div (1) + 28 \times (5 \div 5)^2$
$7 \div 1 + 28 \times (5 \div 5)^2$
$7 \div 1 + 28 \times (1)^2$
$7 \div 1 + 28 \times 1^2$
$7 \div 1 + 28 \times 1$
$7 + 28 \times 1$
$7 + 28$
35

Day 53

① $9^2 \times 2 \div 81 - 1 + 6^2 - 30$
$81 \times 2 \div 81 - 1 + 6^2 - 30$
$81 \times 2 \div 81 - 1 + 36 - 30$
$162 \div 81 - 1 + 36 - 30$
$2 - 1 + 36 - 30$
$1 + 36 - 30$
$37 - 30$
7

② $5 + 7 - 8 \times 3 \div 24 + 2^2$
$5 + 7 - 8 \times 3 \div 24 + 4$
$5 + 7 - 24 \div 24 + 4$
$5 + 7 - 1 + 4$
$12 - 1 + 4$
$11 + 4$
15

③ $13 - (42 - 6 \times 7) + 5 \times 5$
$13 - (42 - 42) + 5 \times 5$
$13 - (0) + 5 \times 5$
$13 - 0 + 5 \times 5$
$13 - 0 + 25$
$13 + 25$
38

④ $18 \div 9 + 5 \times (4 + 4 - 5)^2$
$18 \div 9 + 5 \times (8 - 5)^2$
$18 \div 9 + 5 \times (3)^2$
$18 \div 9 + 5 \times 3^2$
$18 \div 9 + 5 \times 9$
$2 + 5 \times 9$
$2 + 45$
47

⑤ $37 \div (10 + 11 \times 1 + 16 \times 1)$
$37 \div (10 + 11 + 16 \times 1)$
$37 \div (10 + 11 + 16)$
$37 \div (21 + 16)$
$37 \div (37)$
$37 \div 37$
1

⑥ $(2 \times 2)^2 + 5 - 13 \div 13 + 17$
$(4)^2 + 5 - 13 \div 13 + 17$
$4^2 + 5 - 13 \div 13 + 17$
$16 + 5 - 13 \div 13 + 17$
$16 + 5 - 1 + 17$
$21 - 1 + 17$
$20 + 17$
37

⑦ $29 + (7 + 8 - 1 \times 1)^2 \div 98$
$29 + (7 + 8 - 1)^2 \div 98$
$29 + (7 + 7)^2 \div 98$
$29 + (14)^2 \div 98$
$29 + 14^2 \div 98$
$29 + 196 \div 98$
$29 + 2$
31

⑧ $(4 - 3)^2 + (9 \div 9)^2 \times (1 \times 3)$
$(1)^2 + (9 \div 9)^2 \times (1 \times 3)$
$1^2 + (9 \div 9)^2 \times (1 \times 3)$
$1^2 + (1)^2 \times (1 \times 3)$
$1^2 + 1^2 \times (1 \times 3)$
$1^2 + 1^2 \times (3)$
$1^2 + 1^2 \times 3$
$1 + 1^2 \times 3$
$1 + 1 \times 3$
$1 + 3$
4

Day 54

① $(9 + 2)^2 - 6 \div 1 \times (35 - 18)$
$(11)^2 - 6 \div 1 \times (35 - 18)$
$11^2 - 6 \div 1 \times (35 - 18)$
$11^2 - 6 \div 1 \times (17)$
$11^2 - 6 \div 1 \times 17$
$121 - 6 \div 1 \times 17$
$121 - 6 \times 17$
$121 - 102$
19

② $8 \times 5 + (4 - 2)^2 - 2 \div 1$
$8 \times 5 + (2)^2 - 2 \div 1$
$8 \times 5 + 2^2 - 2 \div 1$
$8 \times 5 + 4 - 2 \div 1$
$40 + 4 - 2 \div 1$
$40 + 4 - 2$
$44 - 2$
42

③ $49 \div 7^2 + 3 - 18 \div (4^2 + 2)$
$49 \div 7^2 + 3 - 18 \div (16 + 2)$
$49 \div 7^2 + 3 - 18 \div (18)$
$49 \div 7^2 + 3 - 18 \div 18$
$49 \div 49 + 3 - 18 \div 18$
$1 + 3 - 18 \div 18$
$1 + 3 - 1$
$4 - 1$
3

④ $28 - 5 \times 2 \div 10 + 4 \times 5$
$28 - 10 \div 10 + 4 \times 5$
$28 - 1 + 4 \times 5$
$28 - 1 + 20$
$27 + 20$
47

⑤ $(23 - 24 \div 8 + 10) \times 2 - 5$
$(23 - 3 + 10) \times 2 - 5$
$(20 + 10) \times 2 - 5$
$(30) \times 2 - 5$
$30 \times 2 - 5$
$60 - 5$
55

⑥ $34 + 15 - (2 \times 7 - 1)^2 \div 169$
$34 + 15 - (14 - 1)^2 \div 169$
$34 + 15 - (13)^2 \div 169$
$34 + 15 - 13^2 \div 169$
$34 + 15 - 169 \div 169$
$34 + 15 - 1$
$49 - 1$
48

⑦ $4 \times 7 + 4^2 - 195 \div (1^2 + 4)$
$4 \times 7 + 4^2 - 195 \div (1 + 4)$
$4 \times 7 + 4^2 - 195 \div (5)$
$4 \times 7 + 4^2 - 195 \div 5$
$4 \times 7 + 16 - 195 \div 5$
$28 + 16 - 195 \div 5$
$28 + 16 - 39$
$44 - 39$
5

⑧ $30 \div 2 - 3 \times 2 + 20 \times 1$
$15 - 3 \times 2 + 20 \times 1$
$15 - 6 + 20 \times 1$
$15 - 6 + 20$
$9 + 20$
29

Day 55

① $100 \div 10 + (8 \times 4) - 36 \div 6$
$100 \div 10 + (32) - 36 \div 6$
$100 \div 10 + 32 - 36 \div 6$
$10 + 32 - 36 \div 6$
$10 + 32 - 6$
$42 - 6$
36

② $(150 + 40 \div 8 - 6 + 8) - 157$
$(150 + 5 - 6 + 8) - 157$
$(155 - 6 + 8) - 157$
$(149 + 8) - 157$
$(157) - 157$
$157 - 157$
0

③ $3 \times (2 - 1)^2 + (34 \div 17 - 1^2)$
$3 \times (1)^2 + (34 \div 17 - 1^2)$
$3 \times 1^2 + (34 \div 17 - 1^2)$
$3 \times 1^2 + (34 \div 17 - 1)$
$3 \times 1^2 + (2 - 1)$
$3 \times 1^2 + (1)$
$3 \times 1^2 + 1$
$3 \times 1 + 1$
$3 + 1$
4

④ $(50 - 9 \times 5) - 64 \div 4^2 + 8^2$
$(50 - 45) - 64 \div 4^2 + 8^2$
$(5) - 64 \div 4^2 + 8^2$
$5 - 64 \div 4^2 + 8^2$
$5 - 64 \div 16 + 8^2$
$5 - 64 \div 16 + 64$
$5 - 4 + 64$
$1 + 64$
65

⑤ $9 - 9 + 1 \times 2\ (4 \times 3)^2$
$9 - 9 + 1 \times 2\ (12)^2$
$9 - 9 + 1 \times 2 \times 12^2$
$9 - 9 + 1 \times 2 \times 144$
$9 - 9 + 2 \times 144$
$9 - 9 + 288$
$0 + 288$
288

⑥ $7^2 - (26 + 74) \div 20 \times (7 \div 1)$
$7^2 - (100) \div 20 \times (7 \div 1)$
$7^2 - 100 \div 20 \times (7 \div 1)$
$7^2 - 100 \div 20 \times (7)$
$7^2 - 100 \div 20 \times 7$
$49 - 100 \div 20 \times 7$
$49 - 5 \times 7$
$49 - 35$
14

⑦ $19 + (8 - 8 \div 8 \times 2 - 4)^2$
$19 + (8 - 1 \times 2 - 4)^2$
$19 + (8 - 2 - 4)^2$
$19 + (6 - 4)^2$
$19 + (2)^2$
$19 + 2^2$
$19 + 4$
23

⑧ $5 \times (36 \div 9 + 9 - 12) \times 5$
$5 \times (4 + 9 - 12) \times 5$
$5 \times (13 - 12) \times 5$
$5 \times (1) \times 5$
$5 \times 1 \times 5$
5×5
25

Day 56

① $(50 - 30) \times (2 + 1)^2 - 81 \div 9$
$(20) \times (2 + 1)^2 - 81 \div 9$
$20 \times (2 + 1)^2 - 81 \div 9$
$20 \times (3)^2 - 81 \div 9$
$20 \times 3^2 - 81 \div 9$
$20 \times 9 - 81 \div 9$
$180 - 81 \div 9$
$180 - 9$
171

② $3^2 \div 9 + (5 \times 3 - 15) \times 3$
$3^2 \div 9 + (15 - 15) \times 3$
$3^2 \div 9 + (0) \times 3$
$3^2 \div 9 + 0 \times 3$
$9 \div 9 + 0 \times 3$
$1 + 0 \times 3$
$1 + 0$
1

③ $26 + 2^2 \times 2 - 7^2 \div 49 - 0$
$26 + 4 \times 2 - 7^2 \div 49 - 0$
$26 + 4 \times 2 - 49 \div 49 - 0$
$26 + 8 - 49 \div 49 - 0$
$26 + 8 - 1 - 0$
$34 - 1 - 0$
$33 - 0$
33

④ $5 \times (17 - 25 \div 5) \times 1 + 5$
$5 \times (17 - 5) \times 1 + 5$
$5 \times (12) \times 1 + 5$
$5 \times 12 \times 1 + 5$
$60 \times 1 + 5$
$60 + 5$
65

⑤ $10 + 10 - (100 \div 10) + (1 \times 3)^2$
$10 + 10 - (10) + (1 \times 3)^2$
$10 + 10 - 10 + (1 \times 3)^2$
$10 + 10 - 10 + (3)^2$
$10 + 10 - 10 + 3^2$
$10 + 10 - 10 + 9$
$20 - 10 + 9$
$10 + 9$
19

⑥ $81 \div 3^2 + 25 \div 5^2 \times 18 \div 3$
$81 \div 9 + 25 \div 5^2 \times 18 \div 3$
$81 \div 9 + 25 \div 25 \times 18 \div 3$
$9 + 25 \div 25 \times 18 \div 3$
$9 + 1 \times 18 \div 3$
$9 + 18 \div 3$
$9 + 6$
15

⑦ $(6^2 \div 3) \times (2^2 + 2) - (4 + 3^2)$
$(36 \div 3) \times (2^2 + 2) - (4 + 3^2)$
$(12) \times (2^2 + 2) - (4 + 3^2)$
$12 \times (2^2 + 2) - (4 + 3^2)$
$12 \times (4 + 2) - (4 + 3^2)$
$12 \times (6) - (4 + 3^2)$
$12 \times 6 - (4 + 3^2)$
$12 \times 6 - (4 + 9)$
$12 \times 6 - (13)$
$12 \times 6 - 13$
$72 - 13$
59

⑧ $(15 - 7) \div 8 \times 9 - 9 + 1^2$
$(8) \div 8 \times 9 - 9 + 1^2$
$8 \div 8 \times 9 - 9 + 1^2$
$8 \div 8 \times 9 - 9 + 1$
$1 \times 9 - 9 + 1$
$9 - 9 + 1$
$0 + 1$
1

Day 57

① $4 \times 4^2 - 64 + 9^2 \div 3^2 - 4$
$4 \times 16 - 64 + 9^2 \div 3^2 - 4$
$4 \times 16 - 64 + 81 \div 3^2 - 4$
$4 \times 16 - 64 + 81 \div 9 - 4$
$64 - 64 + 81 \div 9 - 4$
$64 - 64 + 9 - 4$
$0 + 9 - 4$
$9 - 4$
5

② $22 + 23 \times 10^2 \div 100 - 10 + 17$
$22 + 23 \times 100 \div 100 - 10 + 17$
$22 + 2300 \div 100 - 10 + 17$
$22 + 23 - 10 + 17$
$45 - 10 + 17$
$35 + 17$
52

③ $(5 - 5)^2 + (3 \times 3)^2 \div (3 \times 3)$
$(0)^2 + (3 \times 3)^2 \div (3 \times 3)$
$0^2 + (3 \times 3)^2 \div (3 \times 3)$
$0^2 + (9)^2 \div (3 \times 3)$
$0^2 + 9^2 \div (3 \times 3)$
$0^2 + 9^2 \div (9)$
$0^2 + 9^2 \div 9$
$0 + 9^2 \div 9$
$0 + 81 \div 9$
$0 + 9$
9

④ $(36 \div 6 + 8 \times 1) \times (6 \div 6)$
$(6 + 8 \times 1) \times (6 \div 6)$
$(6 + 8) \times (6 \div 6)$
$(14) \times (6 \div 6)$
$14 \times (6 \div 6)$
$14 \times (1)$
14×1
14

⑤ $(7 + 5)^2 - (8 \div 2 + 2 \times 8)$
$(12)^2 - (8 \div 2 + 2 \times 8)$
$12^2 - (8 \div 2 + 2 \times 8)$
$12^2 - (4 + 2 \times 8)$
$12^2 - (4 + 16)$
$12^2 - (20)$
$12^2 - 20$
$144 - 20$
124

⑥ $9 \div (49 - 40) + (3 + 7 \times 2)^2$
$9 \div (9) + (3 + 7 \times 2)^2$
$9 \div 9 + (3 + 7 \times 2)^2$
$9 \div 9 + (3 + 14)^2$
$9 \div 9 + (17)^2$
$9 \div 9 + 17^2$
$9 \div 9 + 289$
$1 + 289$
290

⑦ $29 - 17 + 3 \times 32 \div 96 - 8$
$29 - 17 + 96 \div 96 - 8$
$29 - 17 + 1 - 8$
$12 + 1 - 8$
$13 - 8$
5

⑧ $(4 \times 2 - 2)^2 - (6 - 3)^2 + 36$
$(8 - 2)^2 - (6 - 3)^2 + 36$
$(6)^2 - (6 - 3)^2 + 36$
$6^2 - (6 - 3)^2 + 36$
$6^2 - (3)^2 + 36$
$6^2 - 3^2 + 36$
$36 - 3^2 + 36$
$36 - 9 + 36$
$27 + 36$
63

Day 58

① $(5^2 - 5) \times 2^2 \div 5 + (7^2 \div 7)$
$(25 - 5) \times 2^2 \div 5 + (7^2 \div 7)$
$(20) \times 2^2 \div 5 + (7^2 \div 7)$
$20 \times 2^2 \div 5 + (7^2 \div 7)$
$20 \times 2^2 \div 5 + (49 \div 7)$
$20 \times 2^2 \div 5 + (7)$
$20 \times 2^2 \div 5 + 7$
$20 \times 4 \div 5 + 7$
$80 \div 5 + 7$
$16 + 7$
23

② $(3^2 \times 4) + (5^2 - 3^2 + 23) - 0$
$(9 \times 4) + (5^2 - 3^2 + 23) - 0$
$(36) + (5^2 - 3^2 + 23) - 0$
$36 + (5^2 - 3^2 + 23) - 0$
$36 + (25 - 3^2 + 23) - 0$
$36 + (25 - 9 + 23) - 0$
$36 + (16 + 23) - 0$
$36 + (39) - 0$
$36 + 39 - 0$
$75 - 0$
75

③ $25 \div 5 + 60 - 5 \times 2 + 19$
$5 + 60 - 5 \times 2 + 19$
$5 + 60 - 10 + 19$
$65 - 10 + 19$
$55 + 19$
74

④ $36 + (7 - 3)^2 \times (10 \div 2) \times 1$
$36 + (4)^2 \times (10 \div 2) \times 1$
$36 + 4^2 \times (10 \div 2) \times 1$
$36 + 4^2 \times (5) \times 1$
$36 + 4^2 \times 5 \times 1$
$36 + 16 \times 5 \times 1$
$36 + 80 \times 1$
$36 + 80$
116

⑤ $48 - 6 \times (1 \div 1 + 7 \times 1)$
$48 - 6 \times (1 + 7 \times 1)$
$48 - 6 \times (1 + 7)$
$48 - 6 \times (8)$
$48 - 6 \times 8$
$48 - 48$
0

⑥ $3^2 \times 81 \div 3^2 - 26 + 12 - 1^2$
$9 \times 81 \div 3^2 - 26 + 12 - 1^2$
$9 \times 81 \div 9 - 26 + 12 - 1^2$
$9 \times 81 \div 9 - 26 + 12 - 1$
$729 \div 9 - 26 + 12 - 1$
$81 - 26 + 12 - 1$
$55 + 12 - 1$
$67 - 1$
66

⑦ $6 + (18 - 36 \div 6) + 150 - 50$
$6 + (18 - 6) + 150 - 50$
$6 + (12) + 150 - 50$
$6 + 12 + 150 - 50$
$18 + 150 - 50$
$168 - 50$
118

⑧ $30 \div 15 + 7 + 5 \times 3^2 \div 3$
$30 \div 15 + 7 + 5 \times 9 \div 3$
$2 + 7 + 5 \times 9 \div 3$
$2 + 7 + 45 \div 3$
$2 + 7 + 15$
$9 + 15$
24

Day 59

① $2^2 + (26 - 10) \times 2^2 \div (49 - 41)$
$2^2 + (16) \times 2^2 \div (49 - 41)$
$2^2 + 16 \times 2^2 \div (49 - 41)$
$2^2 + 16 \times 2^2 \div (8)$
$2^2 + 16 \times 2^2 \div 8$
$4 + 16 \times 2^2 \div 8$
$4 + 16 \times 4 \div 8$
$4 + 64 \div 8$
$4 + 8$
12

② $45 - 5 \times 3^2 + 3^2 - 5^2 \div 5$
$45 - 5 \times 9 + 3^2 - 5^2 \div 5$
$45 - 5 \times 9 + 9 - 5^2 \div 5$
$45 - 5 \times 9 + 9 - 25 \div 5$
$45 - 45 + 9 - 25 \div 5$
$45 - 45 + 9 - 5$
$0 + 9 - 5$
$9 - 5$
4

③ $4 \times 8 - (18 \div 6) + 5 - 5$
$4 \times 8 - (3) + 5 - 5$
$4 \times 8 - 3 + 5 - 5$
$32 - 3 + 5 - 5$
$29 + 5 - 5$
$34 - 5$
29

④ $(1 \div 1 \times 7)^2 + 13 - 3 + 6$
$(1 \times 7)^2 + 13 - 3 + 6$
$(7)^2 + 13 - 3 + 6$
$7^2 + 13 - 3 + 6$
$49 + 13 - 3 + 6$
$62 - 3 + 6$
$59 + 6$
65

⑤ $6^2 + 11 - (6 - 6^2 \div 36) \times 3^2$
$6^2 + 11 - (6 - 36 \div 36) \times 3^2$
$6^2 + 11 - (6 - 1) \times 3^2$
$6^2 + 11 - (5) \times 3^2$
$6^2 + 11 - 5 \times 3^2$
$36 + 11 - 5 \times 3^2$
$36 + 11 - 5 \times 9$
$36 + 11 - 45$
$47 - 45$
2

⑥ $15 - (6 \times 2 + 3) \div (5 - 2)$
$15 - (12 + 3) \div (5 - 2)$
$15 - (15) \div (5 - 2)$
$15 - 15 \div (5 - 2)$
$15 - 15 \div (3)$
$15 - 15 \div 3$
$15 - 5$
10

⑦ $(3 \div 3) + (4 + 3 \times 2)^2 + 2$
$(1) + (4 + 3 \times 2)^2 + 2$
$1 + (4 + 3 \times 2)^2 + 2$
$1 + (4 + 6)^2 + 2$
$1 + (10)^2 + 2$
$1 + 10^2 + 2$
$1 + 100 + 2$
$101 + 2$
103

⑧ $(5 - 1) \times 2 + 2 \div (6 - 4)$
$(4) \times 2 + 2 \div (6 - 4)$
$4 \times 2 + 2 \div (6 - 4)$
$4 \times 2 + 2 \div (2)$
$4 \times 2 + 2 \div 2$
$8 + 2 \div 2$
$8 + 1$
9

Day 60

① (35 - 17) x 1 + (50 + 9 ÷ 3)
(18) x 1 + (50 + 9 ÷ 3)
18 x 1 + (50 + 9 ÷ 3)
18 x 1 + (50 + 3)
18 x 1 + (53)
18 x 1 + 53
18 + 53
71

② $(6 + 4 - 2)^2 \div (4^2 + 4^2 - 30)$
$(10 - 2)^2 \div (4^2 + 4^2 - 30)$
$(8)^2 \div (4^2 + 4^2 - 30)$
$8^2 \div (4^2 + 4^2 - 30)$
$8^2 \div (16 + 4^2 - 30)$
$8^2 \div (16 + 16 - 30)$
$8^2 \div (32 - 30)$
$8^2 \div (2)$
$8^2 \div 2$
64 ÷ 2
32

③ 49 ÷ 7 + 11 - 100 ÷ 10 x 1
7 + 11 - 100 ÷ 10 x 1
7 + 11 - 10 x 1
7 + 11 - 10
18 - 10
8

④ 8 x 64 ÷ 4 - 2 x 8 - 24
512 ÷ 4 - 2 x 8 - 24
128 - 2 x 8 - 24
128 - 16 - 24
112 - 24
88

⑤ $3^2 \times (2 - 1 + 2) \times (2 \times 5)$
$3^2 \times (1 + 2) \times (2 \times 5)$
$3^2 \times (3) \times (2 \times 5)$
$3^2 \times 3 \times (2 \times 5)$
$3^2 \times 3 \times (10)$
$3^2 \times 3 \times 10$
9 x 3 x 10
27 x 10
270

⑥ $(20 + 10) \div 5 \times 2 \div (2 - 1)^2$
$(30) \div 5 \times 2 \div (2 - 1)^2$
$30 \div 5 \times 2 \div (2 - 1)^2$
$30 \div 5 \times 2 \div (1)^2$
$30 \div 5 \times 2 \div 1^2$
30 ÷ 5 x 2 ÷ 1
6 x 2 ÷ 1
12 ÷ 1
12

⑦ 32 - 9 ÷ 3 + 21 ÷ 7 - 7
32 - 3 + 21 ÷ 7 - 7
32 - 3 + 3 - 7
29 + 3 - 7
32 - 7
25

⑧ $2 \times 5 - 2^2 + 10 - 4 \times 2^2$
$2 \times 5 - 4 + 10 - 4 \times 2^2$
2 x 5 - 4 + 10 - 4 x 4
10 - 4 + 10 - 4 x 4
10 - 4 + 10 - 16
6 + 10 – 16
16 - 16
0

Day 61

① 10 + (5 - (-8))
10 + (5- -8)
10 + (5 + 8)
10 + (13)
10 + 13
23

② ((8 ÷ (-4)) + (-24))
((8 ÷ -4) + (-24))
((-2) + (-24))
(-2 + (-24))
(-2 + -24)
(-2 - 24)
(-26)
-26

③ ((-3) + (-9 + 3))
(-3 + (-9 + 3))
(-3 + (-6))
(-3 + -6)
(-3 - 6)
(-9)
-9

④ (-7 x (-36 ÷ -6))
(-7 x (6))
(-7 x 6)
(-42)
-42

⑤ ((12 - (-3)) + 6)
((12 - -3) + 6)
((12 + 3) + 6)
((15) + 6)
(15 + 6)
(21)
21

⑥ ((-15 + -7) - 8)
((-15 - 7) - 8)
((-22) - 8)
(-22 - 8)
(-30)
-30

⑦ (((-7) + 2) x 9)
((-7 + 2) x 9)
((-5) x 9)
(-5 x 9)
(-45)
-45

⑧ (20 ÷ -5 + (-35))
(20 ÷ -5 + -35)
(20 ÷ -5 - 35)
(-4 - 35)
(-39)
-39

⑨ (2 x (-24 ÷ 8))
(2 x (-3))
(2 x -3)
(-6)
-6

⑩ ((-6) + (-10) + 8)
(-6 + (-10) + 8)
(-6 + -10 + 8)
(-6 -10 + 8)
(-16 + 8)
(-8)
-8

Day 62

① 3 - ((-5) + 9)
3 - (-5 + 9)
3 - (4)
3 - 4
-1

② ((-3) - 5) + 6
(-3 - 5) + 6
(-8) + 6
-8 + 6
-2

③ ((-7) + (-4)) - 2
(-7 + (-4)) - 2
(-7 + -4) - 2
(-7 - 4) – 2
(-11) – 2
-11 - 2
-13

④ (-3) + ((-8) + 12)
-3 + ((-8) + 12)
-3 + (-8 + 12)
-3 + (4)
-3 + 4
1

⑤ (8 - (-7)) + 5
(8 - -7) + 5
(8 + 7) + 5
(15) + 5
15 + 5
20

⑥ ((-6) + 4) + (-1)
(-6 + 4) + (-1)
(-2) + (-1)
-2 + (-1)
-2 + -1
-2 - 1
-3

⑦ ((-3) + 2) + 9
(-3 + 2) + 9
(-1) + 9
-1 + 9
8

⑧ (5 - (-4)) + 7
(5 - -4) + 7
(5 + 4) + 7
(9) + 7
9 + 7
16

⑨ ((-10) - 4) + (-8)
(-10 - 4) + (-8)
(-14) + (-8)
-14 + (-8)
-14 + -8
-14 - 8
-22

⑩ 3 + ((-5) + 9)
3 + (-5 + 9)
3 + (4)
3 + 4
7

Day 63

① -20 - (-15 + 10)
-20 - (-5)
-20 - -5
-20 + 5
-15

② (35 - 11) - 23
(24) - 23
24 - 23
1

③ (7) + (-4 + 2)
7 + (-4 + 2)
7 + (-2)
7 + -2
7 - 2
5

④ 18 - (41 - (-31))
18 - (41 - -31)
18 - (41 + 31)
18 - (72)
18 - 72
-54

⑤ ((-16) + (-5)) - 6
(-16 + (-5)) - 6
(-16 + -5) - 6
(-16 - 5) - 6
(-21) - 6
-21 - 6
-27

⑥ ((-25) - 13 - (-12))
(-25 - 13 - (-12))
(-25 - 13 - -12)
(-25 - 13 + 12)
(-38 + 12)
(-26)
-26

⑦ (17 + (-7)) - 3
(17 + -7) - 3
(17 - 7) - 3
(10) – 3
10 - 3
7

⑧ ((-28) + (-18) - 8)
(-28 + (-18) - 8)
(-28 + -18 - 8)
(-28 - 18 - 8)
(-46 - 8)
(-54)
-54

⑨ (3 + (-3) - (-2))
(3 + -3 - (-2))
(3 - 3 - (-2))
(3 - 3 - -2)
(3 - 3 + 2)
(0 + 2)
(2)
2

⑩ (24 - (-15) - 13)
(24 - -15 - 13)
(24 + 15 - 13)
(39 - 13)
(26)
26

Day 64

① (-36 + (-16) - (-6))
(-36 + -16 - (-6))
(-36 - 16 - (-6))
(-36 - 16 - -6)
(-36 - 16 + 6)
(-52 + 6)
(-46)
-46

② (5) + (-8 + (-3))
5 + (-8 + (-3))
5 + (-8 + -3)
5 + (-8 - 3)
5 + (-11)
5 + -11
5 - 11
-6

③ 12 - ((-9) - (-7))
12 - (-9 - (-7))
12 - (-9- -7)
12 - (-9 + 7)
12 - (-2)
12 - -2
12 + 2
14

④ 47 + ((-13) + 4)
47 + (-13 + 4)
47 + (-9)
47 + -9
47 - 9
38

⑤ (1 + (-1) - (-2))
(1 + -1 - (-2))
(1 - 1 - (-2))
(1 - 1 - -2)
(1 - 1 + 2)
(0 + 2)
(2)
2

⑥ (6-7+(-9))
(6-7+-9)
(6-7-9)
(-1-9)
(-10)
-10

⑦ -29 + (14 - (-8))
-29 + (14 - -8)
-29 + (14 + 8)
-29 + (22)
-29 + 22
-7

⑧ (9 - (-3)) - 22
(9 - -3) - 22
(9 + 3) - 22
(12) – 22
12 - 22
-10

⑨ ((-20) + 10 + (-40))
(-20 + 10 + (-40))
(-20 + 10 + -40)
(-20 + 10 - 40)
(-10 - 40)
(-50)
-50

⑩ ((-3) + (-9 + 3))
(-3 + (-9 + 3))
(-3 + (-6))
(-3 + -6)
(-3 - 6)
(-9)
-9

Day 65

① ((8 -11) + (-6))
((-3) + (-6))
(-3 + (-6))
(-3 + -6)
(-3 - 6)
(-9)
-9

② 4 - ((-9) -7)
4 - (-9 -7)
4 - (-16)
4 - -16
4 + 16
20

③ ((-6) + (-10) + 8)
(-6 + (-10) + 8)
(-6 + -10 + 8)
(-6 - 10 + 8)
(-16 + 8)
(-8)
-8

④ (5 x ((-81) ÷ 9))
(5 x (-81 ÷ 9))
(5 x (-9))
(5 x -9)
(-45)
-45

⑤ ((8 ÷ (-4)) + (-24))
((8 ÷ -4) + (-24))
((-2) + (-24))
(-2 + (-24))
(-2 + -24)
(-2 - 24)
(-26)
-26

⑥ (-11 + 21) x (-3)
(10) x (-3)
10 x (-3)
10 x -3
-30

⑦ 5 x ((-49) ÷ (-7))
5 x (-49 ÷ (-7))
5 x (-49 ÷ -7)
5 x (7)
5 x 7
35

⑧ ((-12) ÷ 6) + 39
(-12 ÷ 6) + 39
(-2) + 39
-2 + 39
37

⑨ ((10 ÷ (-2)) x 10)
((10 ÷ -2) x 10)
((-5) x 10)
(-5 x 10)
(-50)
-50

⑩ (12 - 4 x (-4))
(12 - 4 x -4)
(12 - -16)
(12 + 16)
(28)
28

Day 66

① ((9 x (-3)) ÷ (-27))
((9 x -3) ÷ (-27))
((-27) ÷ (-27))
(-27 ÷ (-27))
(-27 ÷ -27)
(1)
1

② (4 ÷ (-2) - 36)
(4 ÷ -2 - 36)
(-2 - 36)
(-38)
-38

③ 18 + ((-2) x (-5))
18 + (-2 x (-5))
18 + (-2 x -5)
18 + (10)
18 + 10
28

④ ((7 - (-9) + 6))
((7 - -9 + 6))
((7 + 9 + 6))
((16 + 6))
((22))
(22)
22

⑤ (-6 x 8 ÷ 12)
(-48 ÷ 12)
(-4)
-4

⑥ 11 - (21 + (-3))
11 - (21 + -3)
11 - (21 - 3)
11 - (18)
11 - 18
-7

⑦ (9 ÷ 3 - (-31))
(9 ÷ 3 - -31)
(9 ÷ 3 + 31)
(3 + 31)
(34)
34

⑧ ((-16) ÷ 8) x 2
(-16 ÷ 8) x 2
(-2) x 2
-2 x 2
-4

⑨ ((-26) + 14 - (-14))
(-26 + 14 - (-14))
(-26 + 14 - -14)
(-26 + 14 + 14)
(-12 + 14)
(2)
2

⑩ (3 + (-6) ÷ (-3))
(3 + -6 ÷ (-3))
(3 - 6 ÷ (-3))
(3 - 6 ÷ -3)
(3 - -2)
(3 + 2)
(5)
5

Day 67

① ((24 - 44) x (-4))
((-20) x (-4))
(-20 x (-4))
(-20 x -4)
(80)
80

② (2 x (-24 ÷ 8))
(2 x (-3))
(2 x -3)
(-6)
-6

③ (20 - 40 + (-25))
(20 - 40 + -25)
(20 - 40 - 25)
(-20 - 25)
(-45)
-45

④ ((-81) ÷ (3-2))
(-81 ÷ (3-2))
(-81 ÷ (1))
(-81 ÷ 1)
(-81)
-81

⑤ ((-7) + (-7) -7)
(-7 + (-7) -7)
(-7 + -7 -7)
(-7 - 7 -7)
(-14 -7)
(-21)
-21

⑥ (8 ÷ (-2)) x 2
(8 ÷ -2) x 2
(-4) x 2
-4 x 2
-8

⑦ (5 x 5 + (-5))
(5 x 5 + -5)
(5 x 5 - 5)
(25 - 5)
(20)
20

⑧ (3 x (-36 ÷ 6))
(3 x (-6))
(3 x -6)
(-18)
-18

⑨ ((-42 ÷ 6) + 15)
((-7) + 15)
(-7 + 15)
(8)
8

⑩ (13 - ((-4) - 37))
(13 - (-4 - 37))
(13 - (-41))
(13 - -41)
(13 + 41)
(54)
54

Day 68

① ((-18 ÷ 9) -7)
((-2) -7)
(-2 -7)
(-9)
-9

② (5 x 2 ÷ -10)
(10 ÷ -10)
(-1)
-1

③ (4 + (-3) x -6)
(4 + -3 x -6)
(4 - -18)
(4 + 18)
(22)
22

④ (20 ÷ -5 + (-35))
(20 ÷ -5 + -35)
(20 ÷ -5 - 35)
(-4 - 35)
(-39)
-39

⑤ ((8 - 2) ÷ -6)
((6) ÷ -6)
(6 ÷ -6)
(-1)
-1

⑥ (-7 x (-36 ÷ -6))
(-7 x (6))
(-7 x 6)
(-42)
-42

⑦ (100 ÷ -50 - 10)
(-2 - 10)
(-12)
-12

⑧ ((17 + (-19)) x 2)
((17 + -19) x 2)
((17 - 19) x 2)
((-2) x 2)
(-2 x 2)
(-4)
-4

⑨ ((3 + 9) - (-39))
((12) - (-39))
(12 - (-39))
(12 - -39)
(12 + 39)
(51)
51

⑩ ((-34) + (-13) + 7)
(-34 + (-13) + 7)
(-34 + -13 + 7)
(-34 - 13 + 7)
(-47 + 7)
(-40)
-40

Day 69

① (-1 x 5 x 4)
(-5 x 4)
(-20)
-20

② ((-5) - ((-2) + (-10)))
(-5 - ((-2) + (-10)))
(-5 - (-2 + (-10)))
(-5 - (-2 + -10))
(-5 - (-2 - 10))
(-5 - (-12))
(-5 - -12)
(-5 + 12)
(7)
7

③ ((24 ÷ 6) + (-24))
((4) + (-24))
(4 + (-24))
(4 + -24)
(4 - 24)
(-20)
-20

④ ((-0) - ((-50) ÷ -5))
(0 - ((-50) ÷ -5))
(0 - (-50 ÷ -5))
(0 - (10))
(0 - 10)
(-10)
-10

⑤ ((-15 + -7) - 8)
((-15 - 7) - 8)
((-22) - 8)
(-22 - 8)
(-30)
-30

⑥ (37 + 48 + (-17))
(37 + 48 + -17)
(37 + 48 -17)
(85 -17)
(68)
68

⑦ ((1) ((-14) + (-42)))
(1 ((-14) + (-42)))
(1 (-14 + (-42)))
(1 (-14 + -42))
(1 (-14 - 42))
(1 (-56))
(1 x -56)
-56

⑧ (10 ÷ ((-6) - 4))
(10 ÷ (-6 - 4))
(10 ÷ (-10))
(10 ÷ -10)
(-1)
-1

⑨ ((-11) -6) ÷ -17
(-11 - 6) ÷ -17
(-17) ÷ -17
-17 ÷ -17
1

⑩ (7- (-3) + (-2))
(7- -3 + (-2))
(7 + 3 + (-2))
(7 + 3 + -2)
(7 + 3 - 2)
(10 - 2)
(8)
8

Day 70

① (((-19) - (-17)) + (-15))
((-19 - (-17)) + (-15))
((-19 - -17) + (-15))
((-19 + 17) + (-15))
((-2) + (-15))
(-2 + (-15))
(-2 + -15)
(-2 - 15)
(-17)
-17

② ((8 x -4) ÷ -16)
((-32) ÷ -16)
(-32 ÷ -16)
(-2)
2

③ (3 - 5 x (-6))
(3 - 5 x -6)
(3 - -30)
(3 + 30)
(33)
33

④ (((-7) + 2) x 9)
((-7 + 2) x 9)
((-5) x 9)
(-5 x 9)
(-45)
-45

⑤ ((-1) + ((-6) - 7))
(-1 + ((-6) - 7))
(-1 + (-6 - 7))
(-1 + (-13))
(-1 + -13)
(-1 - 13)
(-14)
-14

⑥ (-40 ÷ 4 x -3)
(-10 x -3)
(30)
30

⑦ ((12 - (-3)) + 6)
((12 - -3) + 6)
((12 + 3) + 6)
((15) + 6)
(15 + 6)
(21)
21

⑧ ((9 x -2) + (-8))
((-18) + (-8))
(-18 + (-8))
(-18 + -8)
(-18 - 8)
(-26)
-26

⑨ ((-0) + ((-29) ÷ 1))
(0 + ((-29) ÷ 1))
(0 + (-29 ÷ 1))
(0 + (-29))
(0 + -29)
(0 - 29)
(-29)
-29

⑩ ((21 ÷ (7)) - (-21))
((21 ÷ 7) - (-21))
((3) - (-21))
(3 - (-21))
(3 - -21)
(3 + 21)
(24)
24

Day 71

1. $\frac{(35 + 45 \div 3)}{5(8 - 6)} = \frac{(35 + 15)}{5(2)} = \frac{(50)}{5 \times 2} = \frac{50}{10} = 5$

2. $\frac{9 - 32 \div 8 - 2}{5^2 - 3 \times 7} = \frac{9 - 4 - 2}{25 - 3 \times 7} = \frac{5 - 2}{25 - 21} = \frac{3}{4}$

3. $\frac{7^2 - 6 \times 8}{6 + 35 \div 7} = \frac{49 - 6 \times 8}{6 + 5} = \frac{49 - 48}{11} = \frac{1}{11}$

4. $\frac{2(40 - 4)}{3^2} = \frac{2(36)}{9} = \frac{2 \times 36}{9} = \frac{72}{9} = 8$

5. $\frac{2(17 - 6)}{60 \div 5 + 5 \times 3} = \frac{2(11)}{12 + 5 \times 3} = \frac{2 \times 11}{12 + 15} = \frac{22}{27}$

6. $\frac{100 - 5^2 \times 3}{4 + 12 \times 2} = \frac{100 - 25 \times 3}{4 + 24} = \frac{100 - 75}{28} = \frac{25}{28}$

7. $\frac{2^2 \times 2}{2(6^2 - 4(5 + 1))} = \frac{4 \times 2}{2(6^2 - 4(6))} = \frac{8}{2(6^2 - 4 \times 6)} = \frac{8}{2(36 - 4 \times 6)} = \frac{8}{2(36 - 24)} = \frac{8}{2(12)} = \frac{8}{2 \times 12} = \frac{8}{24} = \frac{1}{3}$

8. $\frac{11 + 3 \times 9 - 2}{(53 - 8) \div 5} = \frac{11 + 27 - 2}{(45) \div 5} = \frac{38 - 2}{45 \div 5} = \frac{36}{9} = 4$

Day 72

1. $\frac{3^2 + 13}{50 - 5^2} = \frac{9 + 13}{50 - 25} = \frac{22}{25}$

2. $\frac{(22 \div 11 \times 7) + 4}{44 - (6 \times 7)} = \frac{(2 \times 7) + 4}{44 - (42)} = \frac{(14) + 4}{44 - 42} = \frac{14 + 4}{44 - 42} = \frac{18}{2} = 9$

3. $\frac{8^2 \div (7 + 9)}{5 \times 3 - 2^2} = \frac{8^2 \div (16)}{5 \times 3 - 4} = \frac{8^2 \div 16}{15 - 4} = \frac{64 \div 16}{11} = \frac{4}{11}$

4. $\frac{11 - 6 + 6}{28} = \frac{5 + 6}{28} = \frac{11}{28}$

5. $\frac{(13 - 3) + 3^2}{4 \times 7 + 8 - 15} = \frac{(10) + 3^2}{28 + 8 - 15} = \frac{10 + 3^2}{36 - 15} = \frac{10 + 9}{21} = \frac{19}{21}$

6. $\frac{(8^2 + (45 \div 15))}{(39 - 29) \times 7} = \frac{(8^2 + (3))}{(10) \times 7} = \frac{(8^2 + 3)}{10 \times 7} = \frac{(64 + 3)}{70} = \frac{(67)}{70} = \frac{67}{70}$

7. $\frac{(2 \times 5) - (8 \div 4)}{49 \div 7^2 + 7} = \frac{(10) - (8 \div 4)}{49 \div 49 + 7} = \frac{10 - (8 \div 4)}{1 + 7} = \frac{10 - (2)}{1 + 7} = \frac{10 - (2)}{8} = \frac{10 - 2}{8} = \frac{8}{8} = 1$

8. $\frac{24 - 4^2 \div 4 + 30}{15 + (9 \times 4)} = \frac{24 - 16 \div 4 + 30}{15 + (36)} = \frac{24 - 4 + 30}{15 + 36} = \frac{20 + 30}{51} = \frac{50}{51}$

Day 73

1.
$\frac{(20 \times 2) - 35}{9 + 7 \div 7}$
$\frac{(40) - 35}{9 + 1}$
$\frac{40 - 35}{10}$
$\frac{5}{10}$
$\frac{1}{2}$

2.
$\frac{6 + (8 - 4) \times 6}{10 \div 5 + 8}$
$\frac{6 + (4) \times 6}{2 + 8}$
$\frac{6 + 4 \times 6}{10}$
$\frac{6 + 24}{10}$
$\frac{30}{10}$
3

3.
$\frac{36 - 10 + 8^2}{15 \times 3^2 \div 9}$
$\frac{36 - 10 + 64}{15 \times 9 \div 9}$
$\frac{26 + 64}{135 \div 9}$
$\frac{90}{15}$
6

4.
$\frac{4^2 \div 2 \times 5 - 7}{(28 - 25 + 7)^2}$
$\frac{16 \div 2 \times 5 - 7}{(3 + 7)^2}$
$\frac{8 \times 5 - 7}{(10)^2}$
$\frac{40 - 7}{10^2}$
$\frac{33}{100}$

5.
$\frac{9 \times (4 - 2)^2}{(30 \div 15 - 2) + 36}$
$\frac{9 \times (2)^2}{(2 - 2) + 36}$
$\frac{9 \times 2^2}{(0) + 36}$
$\frac{9 \times 4}{0 + 36}$
$\frac{36}{36}$
1

6.
$\frac{(16 - 6)^2 \times 2}{5 \times (24 - 14)}$
$\frac{(10)^2 \times 2}{5 \times (10)}$
$\frac{10^2 \times 2}{5 \times 10}$
$\frac{100 \times 2}{50}$
$\frac{200}{50}$
4

7.
$\frac{(7 + (7 \times 3) \div 21)}{((91 - 10) \div 3^2)}$
$\frac{(7 + (21) \div 21)}{((81) \div 3^2)}$
$\frac{(7 + 21 \div 21)}{(81 \div 3^2)}$
$\frac{(7 + 1)}{(81 \div 9)}$
$\frac{(8)}{(9)}$
$\frac{8}{9}$

8.
$\frac{(30 \div 6 + 2)^2 - 5^2}{3 \times 9 - 1}$
$\frac{(5 + 2)^2 - 5^2}{27 - 1}$
$\frac{(7)^2 - 5^2}{27 - 1}$
$\frac{7^2 - 5^2}{26}$
$\frac{49 - 5^2}{26}$
$\frac{49 - 25}{26}$
$\frac{24}{26}$
$\frac{12}{13}$

Day 74

1.
$\frac{((12 \div 6) \times 3)}{90 \div (1 \times 10)}$
$\frac{((2) \times 3)}{90 \div (10)}$
$\frac{(2 \times 3)}{90 \div 10}$
$\frac{(6)}{9}$
$\frac{6}{9}$
$\frac{2}{3}$

2.
$\frac{25 - 13 + (16 \div 8)}{(15 - 9) \times 2^2}$
$\frac{25 - 13 + (2)}{(6) \times 2^2}$
$\frac{25 - 13 + 2}{6 \times 2^2}$
$\frac{12 + 2}{6 \times 4}$
$\frac{14}{24}$
$\frac{7}{12}$

3.
$\frac{(4 \times 1)^2 \div 2}{5^2 - (8 + 9)}$
$\frac{(4)^2 \div 2}{5^2 - (17)}$
$\frac{4^2 \div 2}{5^2 - 17}$
$\frac{16 \div 2}{25 - 17}$
$\frac{8}{8}$
1

4.
$\frac{18 - 7^2 \div 7 + 3}{8 \times 9 \div 2}$
$\frac{18 - 49 \div 7 + 3}{72 \div 2}$
$\frac{18 - 7 + 3}{36}$
$\frac{11 + 3}{36}$
$\frac{14}{36}$
$\frac{7}{18}$

5.
$\frac{(8 \div 2)^2 - 6}{(8 + 7)2 - 25}$
$\frac{(4)^2 - 6}{(15)2 - 25}$
$\frac{4^2 - 6}{15 \times 2 - 25}$
$\frac{16 - 6}{30 - 25}$
$\frac{10}{5}$
2

6.
$\frac{(3^2 + 16 - 3^2)}{39 - 10 + 4}$
$\frac{(9 + 16 - 3^2)}{29 + 4}$
$\frac{(9 + 16 - 9)}{33}$
$\frac{(25 - 9)}{33}$
$\frac{(16)}{33}$
$\frac{16}{33}$

7.
$\frac{19 \times 3 - 49 \div 7}{9 \times 9 \div 81}$
$\frac{57 - 49 \div 7}{81 \div 81}$
$\frac{57 - 7}{1}$
$\frac{50}{1}$
50

8.
$\frac{12 - 10 + 1^2 \times 4}{(11 - 7)^2 \times 2}$
$\frac{12 - 10 + 1 \times 4}{(4)^2 \times 2}$
$\frac{12 - 10 + 4}{4^2 \times 2}$
$\frac{2 + 4}{16 \times 2}$
$\frac{6}{32}$
$\frac{3}{16}$

Day 75

① $\frac{18 + (5 \times 2)^2}{59}$

$\frac{18 + (10)^2}{59}$

$\frac{18 + 10^2}{59}$

$\frac{18 + 100}{59}$

$\frac{118}{59}$

2

② $\frac{36 \div 12 - 1}{(3 + 2)13}$

$\frac{3 - 1}{(5)13}$

$\frac{2}{5 \times 13}$

$\frac{2}{65}$

③ $\frac{(7^2 + 11)4}{(12 - 10)^2}$

$\frac{(49 + 11)4}{(2)^2}$

$\frac{(60)4}{2^2}$

$\frac{60 \times 4}{4}$

$\frac{240}{4}$

60

④ $\frac{(4 + 5)^2}{(9 - 2) \times (8 + 5)}$

$\frac{(9)^2}{(7) \times (8 + 5)}$

$\frac{9^2}{7 \times (8 + 5)}$

$\frac{81}{7 \times (13)}$

$\frac{81}{7 \times 13}$

$\frac{81}{91}$

⑤ $\frac{6 \times 3 \div 9}{28 \div (4 - 2)^2 + 2}$

$\frac{18 \div 9}{28 \div (2)^2 + 2}$

$\frac{2}{28 \div 4 + 2}$

$\frac{2}{7 + 2}$

$\frac{2}{9}$

⑥ $\frac{(18 - (8 + 6))}{49 \div 7}$

$\frac{(18 - (14))}{7}$

$\frac{(18 - 14)}{7}$

$\frac{(4)}{7}$

$\frac{4}{7}$

⑦ $\frac{8 + ((5 \times 5) - 8)}{11 - (7 + 3)}$

$\frac{8 + ((25) - 8)}{11 - (10)}$

$\frac{8 + (25 - 8)}{11 - 10}$

$\frac{8 + (17)}{1}$

$\frac{8 + 17}{1}$

$\frac{25}{1}$

25

⑧ $\frac{(9^2 - (4 + 2 \times 2)^2)}{5((7 - 3)^2 + 2^2)}$

$\frac{(9^2 - (4 + 4)^2)}{5((4)^2 + 2^2)}$

$\frac{(9^2 - (8)^2)}{5(4^2 + 2^2)}$

$\frac{(9^2 - 8^2)}{5(16 + 2^2)}$

$\frac{(81 - 8^2)}{5(16 + 4)}$

$\frac{(81 - 64)}{5(20)}$

$\frac{(17)}{5 \times 20}$

$\frac{17}{100}$

Day 76

① $\frac{(8 - 6 \div 3)^2}{(2 + 2)^2 \div (2 \times 2)}$

$\frac{(8 - 2)^2}{(4)^2 \div (2 \times 2)}$

$\frac{(6)^2}{4^2 \div (2 \times 2)}$

$\frac{6^2}{4^2 \div (2 \times 2)}$

$\frac{36}{4^2 \div (4)}$

$\frac{36}{4^2 \div 4}$

$\frac{36}{16 \div 4}$

$\frac{36}{4}$

9

② $\frac{((4^2 \times 2) + 7)}{(6^2 + 3)}$

$\frac{((16 \times 2) + 7)}{(36 + 3)}$

$\frac{((32) + 7)}{(39)}$

$\frac{(32 + 7)}{39}$

$\frac{(39)}{39}$

$\frac{39}{39}$

1

③ $\frac{4 \times 5 - 6}{17 + 6 - 9 \div 3}$

$\frac{20 - 6}{17 + 6 - 3}$

$\frac{14}{23 - 3}$

$\frac{7}{10}$

④ $\frac{12 \div 3 + 26}{(2 \times 5)^2 \div 20}$

$\frac{4 + 26}{(10)^2 \div 20}$

$\frac{30}{10^2 \div 20}$

$\frac{30}{100 \div 20}$

$\frac{30}{100 \div 20}$

$\frac{30}{5}$

6

⑤ $\frac{150 - (7 + 4)^2 \times 1}{6^2}$

$\frac{150 - (11)^2 \times 1}{36}$

$\frac{150 - 11^2 \times 1}{36}$

$\frac{150 - 121 \times 1}{36}$

$\frac{150 - 121}{36}$

$\frac{29}{36}$

⑥ $\frac{6(4 - 3)}{3 \div 3(3)}$

$\frac{6(1)}{3 \div 3 \times 3}$

$\frac{6 \times 1}{1 \times 3}$

$\frac{6}{3}$

2

⑦ $\frac{8^2}{8^2 \times 6 \div 96}$

$\frac{64}{64 \times 6 \div 96}$

$\frac{64}{384 \div 96}$

$\frac{64}{4}$

4

⑧ $\frac{22 \div 11 \times 6 - 10}{0 + 11}$

$\frac{2 \times 6 - 10}{11}$

$\frac{12 - 10}{11}$

$\frac{2}{11}$

Day 77

(1)
$$\frac{6^2 - 7 \times 5}{14 \div 7 \times 4^2}$$
$$\frac{36 - 7 \times 5}{14 \div 7 \times 16}$$
$$\frac{36 - 35}{2 \times 16}$$
$$\frac{1}{32}$$

(2)
$$\frac{8 (5) \div 10}{5^2}$$
$$\frac{8 \times 5 \div 10}{25}$$
$$\frac{40 \div 10}{25}$$
$$\frac{4}{25}$$

(3)
$$\frac{((23 + 19) - 34)}{2 \times (16 \div 8)^2}$$
$$\frac{((42) - 34)}{2 \times (2)^2}$$
$$\frac{(42 - 34)}{2 \times 2^2}$$
$$\frac{(8)}{2 \times 4}$$
$$\frac{8}{8}$$
$$1$$

(4)
$$\frac{(24 \times 3) \div 3^2}{96 \div (2^2 \times 6)}$$
$$\frac{(72) \div 3^2}{96 \div (4 \times 6)}$$
$$\frac{72 \div 3^2}{96 \div 24}$$
$$\frac{72 \div 9}{96 \div 24}$$
$$\frac{8}{96 \div 24}$$
$$\frac{8}{4}$$
$$2$$

(5)
$$\frac{(21 \div (2 + 5))}{41 + 80 \div 10}$$
$$\frac{(21 \div (7))}{41 + 8}$$
$$\frac{(21 \div 7)}{49}$$
$$\frac{(3)}{49}$$
$$\frac{3}{49}$$

(6)
$$\frac{(5)1^2}{((25 - 13) + 7)}$$
$$\frac{5 \times 1^2}{((12) + 7)}$$
$$\frac{5 \times 1}{(12 + 7)}$$
$$\frac{5}{(19)}$$
$$\frac{5}{19}$$

(7)
$$\frac{8 \times 2^2 - 7^2 \div 49 + 1}{5^2 - 17}$$
$$\frac{8 \times 4 - 7^2 \div 49 + 1}{25 - 17}$$
$$\frac{8 \times 4 - 49 \div 49 + 1}{8}$$
$$\frac{32 - 49 \div 49 + 1}{8}$$
$$\frac{32 - 1 + 1}{8}$$
$$\frac{31 + 1}{8}$$
$$\frac{32}{8}$$
$$4$$

(8)
$$\frac{(28 - 16) \times 8 - 15}{(6 \div 6 + 8)^2}$$
$$\frac{(12) \times 8 - 15}{(1 + 8)^2}$$
$$\frac{12 \times 8 - 15}{(9)^2}$$
$$\frac{96 - 15}{9^2}$$
$$\frac{81}{81}$$
$$1$$

Day 78

(1)
$$\frac{21 + (9 \div 3)}{(16 - 13)4}$$
$$\frac{21 + (3)}{(3)4}$$
$$\frac{21 + 3}{3 \times 4}$$
$$\frac{24}{12}$$
$$2$$

(2)
$$\frac{(6 \div 3 + 2 - 2)^2}{(5 - 4 \times 1)^2}$$
$$\frac{(2 + 2 - 2)^2}{(5 - 4)^2}$$
$$\frac{(4 - 2)^2}{(1)^2}$$
$$\frac{(2)^2}{1^2}$$
$$\frac{2^2}{1}$$
$$\frac{4}{1}$$
$$4$$

(3)
$$\frac{38 + 23 - 7 \times 8}{(8)(5)}$$
$$\frac{38 + 23 - 56}{8(5)}$$
$$\frac{61 - 56}{8 \times 5}$$
$$\frac{5}{40}$$
$$\frac{1}{8}$$

(4)
$$\frac{20^2 \div 100}{42 + 8^2 - 24}$$
$$\frac{400 \div 100}{42 + 64 - 24}$$
$$\frac{4}{106 - 24}$$
$$\frac{4}{82}$$
$$\frac{2}{41}$$

(5)
$$\frac{2^2 \times (4^2 + (8 \div 2))}{((9 \div 3) + 3^2) - 2^2}$$
$$\frac{2^2 \times (4^2 + (4))}{((3) + 3^2) - 2^2}$$
$$\frac{2^2 \times (4^2 + 4)}{(3 + 3^2) - 2^2}$$
$$\frac{2^2 \times (16 + 4)}{(3 + 9) - 2^2}$$
$$\frac{2^2 \times (20)}{(12) - 2^2}$$
$$\frac{2^2 \times 20}{12 - 2^2}$$
$$\frac{4 \times 20}{12 - 4}$$
$$\frac{80}{8}$$
$$10$$

(6)
$$\frac{39 + 7^2 - 49}{9^2 \div 3 + 27}$$
$$\frac{39 + 49 - 49}{81 \div 3 + 27}$$
$$\frac{88 - 49}{27 + 27}$$
$$\frac{39}{54}$$
$$\frac{13}{18}$$

(7)
$$\frac{8^2 \times 5}{32}$$
$$\frac{64 \times 5}{32}$$
$$\frac{320}{32}$$
$$10$$

(8)
$$\frac{(12 - 6) + (12 - 6)}{15 \times 2 + 5}$$
$$\frac{(6) + (12 - 6)}{30 + 5}$$
$$\frac{6 + (12 - 6)}{35}$$
$$\frac{6 + (6)}{35}$$
$$\frac{6 + 6}{35}$$
$$\frac{12}{35}$$

Day 79

1.
$\frac{108 \div 6^2 - 1}{3^2 - 5 + 16}$
$\frac{108 \div 36 - 1}{9 - 5 + 16}$
$\frac{3 - 1}{4 + 16}$
$\frac{2}{20}$
$\frac{1}{10}$

2.
$\frac{3 \times 5 + 2}{((2 + 2)^2 - 8) \times 3}$
$\frac{15 + 2}{((4)^2 - 8) \times 3}$
$\frac{17}{(4^2 - 8) \times 3}$
$\frac{17}{(16 - 8) \times 3}$
$\frac{17}{(8) \times 3}$
$\frac{17}{8 \times 3}$
$\frac{17}{24}$

3.
$\frac{(4^2 + 11) \div 3^2}{8 \times 6^2 - 9^2}$
$\frac{(16 + 11) \div 3^2}{8 \times 36 - 9^2}$
$\frac{(27) \div 3^2}{8 \times 36 - 81}$
$\frac{27 \div 3^2}{288 - 81}$
$\frac{27 \div 9}{207}$
$\frac{3}{207}$
$\frac{1}{69}$

4.
$\frac{38 - 33 + 4}{72 \div 8}$
$\frac{5 + 4}{9}$
$\frac{9}{9}$
1

5.
$\frac{((3 \times 8) - 15)}{(24 - (10 \div 5) + 11)}$
$\frac{((24) - 15)}{(24 - (2) + 11)}$
$\frac{(24 - 15)}{(24 - 2 + 11)}$
$\frac{(9)}{(22 + 11)}$
$\frac{9}{(33)}$
$\frac{9}{33}$
$\frac{3}{11}$

6.
$\frac{20 \div 4 + 20}{(15 - 10)^2 \times 5}$
$\frac{5 + 20}{(5)^2 \times 5}$
$\frac{5 + 20}{5^2 \times 5}$
$\frac{25}{25 \times 5}$
$\frac{25}{125}$
$\frac{1}{25}$

7.
$\frac{(23 - 13) \times 6 \div 2}{5 \times 2}$
$\frac{(10) \times 6 \div 2}{10}$
$\frac{10 \times 6 \div 2}{10}$
$\frac{60 \div 2}{10}$
$\frac{30}{10}$
3

8.
$\frac{1(13 - 3 + 5)}{(18 \div 2)3}$
$\frac{1(10 + 5)}{(9)3}$
$\frac{1(15)}{9 \times 3}$
$\frac{1 \times 15}{27}$
$\frac{15}{27}$

Day 80

1.
$\frac{9 - 2 \times 4}{14 \div 7 \times 4^2}$
$\frac{9 - 8}{14 \div 7 \times 16}$
$\frac{1}{2 \times 16}$
$\frac{1}{32}$

2.
$\frac{(3 + 3) \div 1 - 2}{5^2 \div 5 + 3}$
$\frac{(6) \div 1 - 2}{25 \div 5 + 3}$
$\frac{6 \div 1 - 2}{5 + 3}$
$\frac{6 - 2}{8}$
$\frac{4}{8}$
$\frac{1}{2}$

3.
$\frac{10 + (9 - 5)}{21 \times 3}$
$\frac{10 + (4)}{63}$
$\frac{10 + 4}{63}$
$\frac{14}{63}$
$\frac{2}{9}$

4.
$\frac{19 + 13 - 11}{3^2 \times (6 + 4)}$
$\frac{32 - 11}{3^2 \times (10)}$
$\frac{21}{3^2 \times 10}$
$\frac{21}{9 \times 10}$
$\frac{21}{90}$
$\frac{7}{30}$

5.
$\frac{4^2 + 7^2}{74 + 8 \div 2 - 5}$
$\frac{16 + 7^2}{74 + 4 - 5}$
$\frac{16 + 49}{78 - 5}$
$\frac{65}{73}$

6.
$\frac{(5 + 1) \times (11 - 9)}{10 - 3 + 8}$
$\frac{(6) \times (11 - 9)}{7 + 8}$
$\frac{6 \times (11 - 9)}{15}$
$\frac{6 \times (2)}{15}$
$\frac{6 \times 2}{15}$
$\frac{12}{15}$

7.
$\frac{8 \times 2^2 - 14}{3^2}$
$\frac{8 \times 4 - 14}{9}$
$\frac{32 - 14}{9}$
$\frac{18}{9}$
2

8.
$\frac{3 + (4 \times 4) \div 2}{28 \div 7 + 10}$
$\frac{3 + (16) \div 2}{4 + 10}$
$\frac{3 + 16 \div 2}{4 + 10}$
$\frac{3 + 8}{4 + 10}$
$\frac{11}{4 + 10}$
$\frac{11}{14}$

Printed in Germany
by Amazon Distribution
GmbH, Leipzig